Decentralized Governance of Adaptation to Climate Change in Africa

Decentralized Governance of Adaptation to Climate Change in Africa

Esbern Friis-Hansen

CABI

CABI is a trading name of CAB International

CABI
Nosworthy Way
Wallingford
Oxfordshire OX10 8DE
UK

Tel: +44 (0)1491 832111
Fax: +44 (0)1491 833508
E-mail: info@cabi.org
Website: www.cabi.org

CABI
745 Atlantic Avenue
8th Floor
Boston, MA 02111
USA

Tel: +1 (617)682-9015
E-mail: cabi-nao@cabi.org

A catalogue record for this book is available from the British Library, London, UK.

Library of Congress Cataloging-in-Publication Data

Names: Friis-Hansen, Esbern, author, editor.
Title: Decentralized governance of adaptation to climate change in Africa / Esbern Friis-Hansen.
Description: Boston, MA : CABI, [2017] | Includes bibliographical references and index.
Identifiers: LCCN 2017004388 (print) | LCCN 2017008646 (ebook) | ISBN 9781786390769 (hbk : alk. paper) | ISBN 9781786390776 (ePDF) | ISBN 9781786390783 (ePub)
Subjects: LCSH: Climatic changes--Government policy--Africa. | Decentralization in government--Africa.
Classification: LCC QC903.2.A35 F75 2017 (print) | LCC QC903.2.A35 (ebook) | DDC 363.738/7456096--dc23
LC record available at https://lccn.loc.gov/2017004388

ISBN-13: 9781786390769

Commissioning editor: Alex Hollingswoth/David Hemming
Editorial assistant: Emma McCann
Production editor: James Bishop

Typeset by SPi, Pondicherry, India
Printed and bound in the UK by CPI Group (UK) Ltd, Croydon, CR0 4YY

Contents

List of Contributors

Charles Aben, College of Agricultural and Environmental Sciences, Makerere University, Uganda. E-mail: charlesaben@gmail.com

Jacob Agea, College of Agricultural and Environmental Sciences, Makerere University, Uganda. E-mail: jgagea@gmail.com

Bernard Bashaasha, College of Agricultural and Environmental Sciences, Makerere University. E-mail: bashaasha@caes.mak.ac.ug

Sarah Ann Lise D'haen, Institute for Geosciences and Natural Resource Management, University of Copenhagen; Consultant, Climate Analytics, Berlin; and IRI-THESys, Humboldt University, Berlin. E-mail: sarah.dhaen@gmail.com

Esbern Friis-Hansen, Danish Institute for International Studies, Copenhagen, Denmark. E-mail: efh@diis.dk

Mikkel Funder, Danish Institute for International Studies, Copenhagen, Denmark. E-mail: mfu@diis.dk

Carol Mweemba, PhD student, University of Zambia, Zambia. E-mail: carolmweemba@yahoo.com

Isaac Nakendo, Research assistant, National Agricultural Advisory Services, Uganda. E-mail: nakendoaa@yahoo.com

Jonas Østergaard Nielsen, Professor, IRI THESys, Humboldt University, Berlin. E-mail: jonas.ostergaard.nielsen@hu-berlin.de

Imasiku Nyambe, Professor, University of Zambia, Zambia. E-mail: inyambe@unza.zm and inyambe@gmail.com

J.J. Okiror, Lecturer, College of Agricultural and Environmental Sciences, Makerere University, Kampala, Uganda. E-mail: okiror2010@gmail.com

Julie Fogt Rasmussen, Student, Danish Institute for International Studies, Copenhagen, Denmark. E-mail: jufr@diis.dk

Godfrey Suubi, Socio-economist, National Agricultural Research Institute, Uganda. E-mail: suubigodfrey@gmail.com

Foreword

Climate change hazards are a real part of the rural people's life in Africa. Gradual climate change increases variability of weather patterns and reduces the predictability of agricultural seasons, whereas extreme climate change events take the form of severe floods or prolonged droughts. While everyone is affected, the poor are more vulnerable and suffer most from climate change. Autonomous adaptation by individual households is largely inadequate and there is an urgent need for collective adaptive solutions planned and implemented through inclusive governance at the level of climate change action.

Understanding the institutional conditions and dynamics of how subnational governments engage with climate change is crucial because local governments are best situated to implement international and national policies and programmes in practice. Yet, very few fieldwork-based studies have examined how subnational institutions respond to climate change. This book addresses this knowledge gap. Most chapters in this book are an outcome of the Climate Change and Rural Institution (CCRI) studies that were undertaken 2012–2016 with the aim to examine how a range of subnational institutions respond to climate change in Africa and Asia. Although programme findings have been presented and published elsewhere (www.diis.dk; Christoplos *et al.*, 2016), this book presents and reflects on CCRI research findings from Africa.

I am grateful to the Danish Research Council for Development Research for funding the research programme. The Danish Institute for International Studies has provided an excellent institutional and academic environment for undertaking the research. The two CCRI partner institutions in Africa, Makerere University in Kampala and Zambia University in Lusaka, have been very actively engaged in undertaking the research, making it a truly collaborative study. The authors of this book are in debt to the many people who in various ways assisted and contributed to the 11 district-based case studies.

The book comes at a highly appropriate time, as the international community after long negotiations reached an agreement at COP 21 in Paris 2015 for a roadmap for financing support for climate change adaptation. Focus has now shifted on how to take action and implement the agreement in practice. However, only a few national and international programmes have yet to engage with the subnational government level in Africa in a serious way. It is the hope that this book can contribute to enhance our understanding of the institutional dynamics of how local governments respond to climate change and that can be useful for future support for creating an enabling environment for rural climate change adaptation in Africa.

Esbern Friis-Hansen
Senior researcher
Danish Institute for International Studies

1 Introduction

Esbern Friis-Hansen

Towards Implementation of Adaptation to Climate Change in Rural Africa

Climate change is widely acknowledged to be one of the fundamental challenges facing Africa and the entire world in the 21st century. In its fifth assessment, the United Nations Intergovernmental Panel on Climate Change (IPCC) presents overwhelming scientific evidence that human-induced emissions of greenhouse gases are changing the global climate (IPCC, 2013). Climate change mitigation of CO_2 emissions is a global challenge, and the narratives of how to conceptualize climate change mitigation and adaptation and how to act on this knowledge are being negotiated in the Conference of the Parties (COP) under the United Nations Framework Convention on Climate Change (UNFCCC).

During the 21st COP held in Paris in 2015, global agreement was reached on a new clear roadmap for reaching US$100 billion of international finance for climate change mitigation and adaptation for developing countries by 2020. Influenced by the UNFCCC process, many African countries have recently (2012–2013) developed National Climate Change Policies (NCCPs) that set out policy frameworks for the implementation of climate change mitigation and adaptation nationally. NCCPs have centralized governance structures, and international and national support for rural climate change adaptation has until now been limited compared with support to urban centres.

With increasing financial resources on track to become available for climate change adaptation, the attention is now on moving from the COP 21 agreement and National Climate Change Policies to action at the subnational and community levels. Engagement with subnational governments in support of climate change adaptation involves more than capacity-building and finance: it requires a strong focus on multiple governance levels for climate change adaptation, particularly the role of subnational institutions (Nalau *et al.*, 2015; Vogel and Henstra, 2015; Christoplos *et al.*, 2016), and on how support for climate change adaptation can be provided at the level of those who are vulnerable to climate change hazards (Conway and Mustelin, 2014; Lesnikowski *et al.*, 2015).

Yet our understanding of how subnational institutions are responding to climate change, and how they interact with the central state and local communities in practice, is limited. Climate Change and Rural Institutions (CCRI) is a multi-partner research programme (2012–2016) being coordinated by the Danish Institute for International Studies (DIIS) in collaboration with partners in Nepal, Vietnam, Uganda and Zambia. The programme examines how subnational institutions are responding to climate change in

Africa and Asia (Christoplos *et al.*, 2016). This book builds on results from the CCRI research programmes from Africa, supplemented with complementing case studies from Tanzania and Ghana.

A crucial focus throughout this book is the extent to which actors and institutions are relevant for implementing rural climate change adaptation and to which the voices of those who are vulnerable to climate change hazards and accountability towards them are being included in these governance arrangements.

How national policies are translated into action depends on the extent to which the sector concerned is part of the decentralization. Many barriers for action occur when there is an insufficient balance between the devolution of responsibility and the devolution of fiscal and human resources. It is important to gain a deeper understanding of how subnational institutions respond to the challenge of climate change at the 'meso-level' between the central state and local communities. There is a disconnect, and an absence of administrative guidelines, regarding the level of devolution within the climate change policy arena, which again leaves a space for considerable ambiguity over how national climate change policies are to be implemented when, as promised, they will be financed by international communities. Because of the lack of clarity about governance mechanisms in NCCPs and the international pledges of significant financial support for climate change, there are widespread expectations among local government politicians and technical staff that local government is to be a key player in facilitating an enabling environment that allows local communities to adapt to climate change. These expectations have contrasted with the experiences of international climate investments until now, where central ministries have shown themselves eager to exert centralized control over political and fiscal resources. The ambiguity over how to deal with climate change and over the administrative level that should have the mandate to do so has turned climate change into a new arena for both policy-making and intra-government institutional infighting over fiscal resources, authority and legitimacy.

Vogel and Henstra (2015) emphasize that the roles and responsibilities assigned to local government in decentralization reforms are well suited to subnational levels of government, which can play a central role in providing services that permit local climate change adaptation. However, although the challenge of enabling local citizens to adapt to climate change is well suited to local government, as discussed in Chapter 2, national climate change policies in Africa seem inclined to favour a highly centralized and project-based mode of implementation, which until now has failed to address the decentralization process at all seriously.

Meso-level Actors and Climate Change

This study will focus on the contribution that local government has made to the implementation of programmes for climate change adaptation in rural Africa. Climate change adaptation needs institutional anchoring at the local level, in local communities, if national policies are to succeed. The study reviews relevant experiences gained in this field by international agencies, national governments and international NGOs.

Local governments in Africa are mundane and bureaucratic subnational institutions that carry out everyday administration and service delivery. They are very ordinary and may be considered insignificant for international climate change forums. There is, however, a slow but growing recognition within such forums, such as the IPCC and COP, that it is not enough to rely on Central governments alone to implement climate change actions. In recent status reports from the IPCC, the emphasis is on the role of local governments in addressing climate change.

Although international support for climate change in Africa is beginning to materialize, very little of this support has reached subnational government institutions. In spite of this lack of financial support, local governments have often engaged in real actions to deal with climate change. Incentives to act on the part of locally elected councillors include the largely popularly driven pressure of their constituents, who are frequently confronted with climate change hazards such as extreme floods and droughts.

Now that national climate change policies have been developed across Africa, international attention is increasingly turning to how these policies can actually be implemented. While there is a great deal of ambiguity about the content of climate change adaptation, attention is shifting to how support for adaptive action can be implemented and which administrative units should take decisions and control access to funds.

District and subdistrict local governments make up the institutional 'middle ground' between the central state and the local community. Local governments have been thus uniquely positioned to support the adaptation efforts of individual citizens, as well as to mediate policy and funding processes between local and national levels (Funder *et al.*, 2014).

For the past decade, the world's mega-cities have been a central driver of climate change adaptation, are well organized globally, and have had significant influence on the international climate change agenda. As yet, however, local governments in rural areas are not playing any significant role as representatives of this political and administrative level or as actors in their own right. In part, they are an extremely varied group with regard to the degree of fiscal and human resource devolution they enjoy, as well as the extent to which they are democratically representative and downwardly accountable (Funder *et al.*, 2014).

Decentralization reform has dominated most African countries in the past two decades, and in most countries a range of basic services have been devolved by law to subnational political and administrative units such as local government. Decentralization can take three forms: devolution, deconcentration and delegation. In practice, decentralization reforms typically involve a mixture of all three forms across sectors. In its 'true' or full form, decentralization is devolution, which consists of 'central government's transfer of administrative and financial decision-making authority to local governments that have clear and legally recognized jurisdictions within which they provide public services to constituents they are accountable to' (Yilmaz *et al.*, 2008, p.1). Under the deconcentration form of decentralization, local government staff come under the central parent ministry in respect of their salaries, professional conduct and employment careers, and to the chief administrative officer (CAO) for their day-to-day work. Fiscal resources are also typically controlled by the parent ministry in the case of deconcentration. The delegation model is the weakest form of decentralization, where specific tasks or services are delegated to local governments by seconding staff or implementing projects. In this model, local government serves the role of agent rather than partner.

In theory, devolution creates stronger pressures on government performance by improving both upward and downward accountability. 'Decentralization reshapes power relations among the local residents, local governments, producers of local government services, and higher levels of government (including central government). It sets new rules of the political game, helping new local leaders to emerge in the political competition. It thus redefines the interactions between local leaders and their constituencies' (Yilmaz *et al.*, 2008). Devolution can therefore potentially have a considerable positive impact on improving access to service delivery and on its quality. The underlying explanation for such improvements to services is that decentralization shifts the interactions and contractual relationships between local governments (as the providers and producers of services) on the one hand and communities and non-governmental organizations on the other.

The level of devolution of political, fiscal and human resources varies between sectors. Primary education, primary health care and municipal roads typically make up the bulk of the local government budget. In many African countries, natural resource management and agricultural sector services have also been devolved to local government.

It can, however, be a long road from theory to practice, and the success of decentralization in Africa has been mixed, for example, not necessarily leading to greater accountability or greater access to and quality of services (Yilmaz *et al.*, 2008). The explanations for these mixed results are many, including inadequate human resource capacity and poorly functioning political processes. In part, this mixed result can be understood as an effect of incomplete or skewed decentralization reforms, as when responsibilities are devolved to local government, but without a similar devolution of human resources and/or fiscal resources. Another major problem with some local governments is the poor level of downward accountability to citizens in-between elections.

There are three dimensions to devolution in decentralization reforms that provide local government with a discretionary space: political, administrative and fiscal. In most decentralization reforms, a process of devolution that provides a larger discretionary space for local government is always followed by more upward public accountability mechanisms to safeguard against the misuse and abuse of this local discretion. However, legal and institutionalized reforms that provide downward social accountability mechanisms enabling citizens to have the ability and opportunity to demand accountability are not as frequent. Recent research indicates that downward institutional accountability mechanisms that provide local governments with the means and incentives to respond to citizens' demands result in better accountability and better service delivery (Yilmaz *et al.*, 2008; Friis-Hansen and Kyed, 2009).

Methodology

Most of the authors in this book have taken part in the CCRI programme coordinated by DIIS and supported by the Danish International Development Agency (DANIDA). The CCRI programme took place in the period from 2012 to 2016 in collaboration with partners in Uganda and Zambia. Africa (Uganda and Zambia) as well as Asia (Vietnam and Nepal). With the books focus on Sub Saharan Africa, two additional fieldwork countries were added (Tanzania and Ghana). The book draw empirical evidence from ten rural local governments. These were chosen using the following three criteria: (i) they each represent contexts in which a range of climate-related hazards exist and constitute a significant development concern; (ii) they represent shifting vulnerabilities, with prevailing livelihood systems being undermined by gradual climate change, or situations in which increasingly recurrent extreme climate events are posing significant challenges to rural institutions; and (iii) contrasting institutional settings and themes have been chosen to represent differences in terms of the underlying governance patterns and social orders that may influence institutional responses.

The primary data for this book have been collected through qualitative data collection in the field, face-to-face interviews, mainly semi-structured, and focus-group discussions with key stakeholders. These stakeholders range from high political officials to the local community level, including local government technical staff, district council politicians, opinion formers, staff from non-governmental organizations, civil society and local community organizations, and different categories of rural citizens, both the poor and the better off, men and women, the young, etc. The interviews were supported by frequent field visits that allowed for participant observation and a deeper understanding of climate change effects and the adaptation measures taken in individual countries. In many cases follow-up interviews were conducted to clarify selected issues and explore them further. These follow-up interviews, along with the triangulation of different methods, have ensured the validity of the collected empirical material. In some cases, qualitative interviews were recorded, transcribed and coded using NVivo qualitative data analysis software. Desk-based reviews have been undertaken to complement empirical fieldwork, including comprehensive literature searches on climate change and local government policies, implementation frameworks, and legal acts within the respective countries, as well as on both the regional and international levels.

Outline of the Book

The book comprises 13 chapters divided into four sections. The first section explores the cross-sectorial aspects of climate change governance, with a focus on the disconnect between the central and subnational levels of governance and on how climate change has become a contested policy arena. Section 2 explores three different empirical experiences by engaging with the subnational level of governance and using a central government-governed approach, an international non-governmental organization (NGO)-facilitated approach, and a local government-driven approach. Section 3 takes a political ecology view of climate change adaptation in rural areas and seeks to understand the dynamics of such adaptation where governance structures are weak or absent. Section 4 discusses two new trends that are under-explored in the climate change literature, namely mobility and regional agricultural trade. Finally a concluding chapter seeks to draw conclusions across the rich empirical material in the case studies.

The first section of the book addresses the governance of climate change adaptation at the global, national and subnational levels, with an emphasis on the last of these. The section examines how the formulation of NCCPs in Africa is influenced by the global climate change narrative and funding opportunities. It further seeks to understand the government mechanisms associated with NCCP and current attempts to operationalize its implementation at the subnational level. The ambiguous understanding of government officials at multiple levels of governance is giving rise to widespread uncertainties over the mandates, roles and responsibilities for implementing climate change action and providing space and fuel for contestations over authority between the national and subnational administrative and political levels.

Chapter 2, *Implementing African National Climate Change Policies* by Esbern Friis-Hansen, aims to identify the drivers behind climate change policies, the intentions connected with their implementation, and the balance between the national and subnational administrative levels with regard to control over finance, decision making and action. International climate change forums such as the UNFCCC have agreed on a joint policy framework for climate change mitigation and adaptation. This chapter first traces the links between the international climate change agenda and the formulation of NCCPs in Africa. It then goes on to examine the governance mechanisms associated with national climate change policies in Africa and their implementation in practice, particularly the role of subnational institutions (Nalau *et al.*, 2015; Vogel and Henstra, 2015; Christoplos *et al.*, 2016). The chapter then analyses rural climate change adaptation using the principle of subsidiarity. It argues that support for climate change action is likely to be ineffective when fiscal resources and decision making are disproportionally located within central administrations that are not ideally placed to deliver an enabling political and technical environment in which rural communities can adapt to climate change hazards, particularly when they lack the human capacity to do so. The chapter ends by assessing the balance between decision making and implementing power as this affects national and subnational government institutions, respectively.

Chapter 3, *Contemplating Climate Change at Local Government: On-the-ground Politics of Adaptation Delivery in Tanzania* by Sarah Ann Lise D'haen and Jonas Østergaard Nielsen, explores how Tanzania's National Climate Change Strategy (CCS) is one of operationalization at the subnational level. The discourses of adaptation are traced across, within and between different departments and tiers of the Tanzanian government at the subnational levels. A particular focus is how tasks and responsibilities in delivering adaptation within local governments are organized.

On the basis of qualitative interviews with civil servants about their discourses on climate change issues, Chapter 3 examines whether particular discourses can be linked to particular actors, their understandings, interests and (struggles for) power or influence, and, if so, which ones. The chapter focuses on the role of local government in national adaptation action plans and shows how local government bureaucrats navigate and contemplate their mandate in the context of the national adaptation agenda. The chapter also reflects critically on these political processes of adaptation at the local government level in Tanzania and relates them to wider trends, processes and implications of adaptation politics in developing countries. The analysis in the chapter seeks to generate a deeper understanding of the social processes and struggles over authority that are taking place as part of the implementation of climate change strategies at the subnational level in Tanzania.

In Chapter 4, *Climate Change Adaptation and the Politics of Decentralization: the Case of Local Governments in Rural Zambia* by Mikkel Funder, Carol Mweemba and Imasiku Nyambe, the focus is on the importance of rural local governments in Africa as important players in implementing climate change adaptation. Although African cities continue to grow at a faster rate than rural areas, a large proportion of Africa's population continues to live and sustain their livelihoods in rural settings, where many are dependent on the use of natural resources and are thus highly vulnerable to climate change hazards. International and national support for climate change adaptation in rural areas has until now been limited compared with that in urban centres. Most African countries have undergone decentralization reforms implemented over the past two decades, and the administrative and political structures are in place to allow rural local governments to create an enabling environment for adaptation and support to specific activities.

Chapter 4 thus examines the 'real-world context' in which rural local governments in African countries operate in terms of climate adaptation, showing the conditions and constraints they face, and how they respond. The chapter further contributes to our understanding of these issues through a discussion of selected findings from two rural districts in Zambia, drawing on research conducted in 2012–2016 as part of the research programme on Climate Change and Rural Institutions (Christoplos *et al.*, 2016). The focus is less on specific capacity and funding issues, and more on the underlying institutional and political dynamics of which rural local governments form a part. In particular, the chapter discusses how the relationship between the central state and local governments and the associated politics of decentralization influence subnational climate change adaptation.

The second section of the book is concerned with the extent to which the climate change agenda is compliant with the decentralization reforms of the past two decades, and in particular how local governments are engaging with implementing climate change adaptation. The section presents three very different empirical scenarios that each represents a particular current trend. The first is the implementation of centrally governed but decentralized projects. It represents one end of the spectrum, with a centralized project implementation unit and implementation through project institutions operating in parallel with local government institutions, thus being least compliant with decentralization reforms. The project model is an old-fashioned, out-dated and much criticized model for development interventions, but at the same time it is representative of the current trend for how the Global Green Fund and other UN and development agencies involve themselves in support for climate change adaptation. The second trend is implementation facilitated by an international NGO using a rights-based approach that simultaneously supports communities in formulating climate change action plans and building the capacity within local government to include these plans with the district development plans and to request funding from central government. The third trend consists of a local government authority politically and administratively addressing climate change adaptation on its own within the local government legal framework and without any financial and human resource support from central government or international donor agencies.

Chapter 5, *White Elephant in a Changing Climate: a Territorial Approach to Climate Change Adaptation in Uganda* by J.J. Okiror, Esbern Friis-Hansen, Bernard Bashaasha and Isaac Nakendo, examines how a UN-financed project implemented by the Ugandan Ministry of Water and Environment engages with local government. The project reflects a common approach among internationally financed projects, namely the *planned climate change adaptation approach* that seeks to draw expertise from science and that uses external consultants to consult the population and formulate the climate change plan of action. The chapter reveals that the project was implemented using parallel structures to those of local government, which assigned local government staff and elected politicians roles as implementing agents rather than as partners. The chapter analyses the effects of such a model in creating a perception of ownership among local government staff and politicians and in ensuring the project's institutional sustainability.

Chapter 6, *Creating Political Space for Climate Change Adaptation in Northern Ghana* by Julie Fogt Rasmussen and Esbern Friis-Hansen, explores how an international NGO (CARE Denmark) and its national sister organization (CARE Ghana) can become significant players in forging downward accountability links between local communities and local government. Through its Adaptation Learning Programme (ALP), implemented in four African countries since 2010, the programme seeks to create new local political spaces where rural people can discuss the climate change hazards they face and agree on possible adaptive action.

Climate change is a problem that spans multiple scales from the local to the global spheres, making it a challenge across scale. The chapter focuses on how the ALP seeks to create spaces for local-level participation in practice and how it addresses those at whom the participatory approaches are aimed. This chapter aims at revealing the role that NGOs can play in reconfiguring this political space and creating new spaces by showing how participatory approaches are integrated in and received by local communities and local governance in the context of enabling sustainable multi-level and cross-scale governance structures for climate change adaptation.

Chapter 7, *Local Politics of Climate Change Adaptation in Uganda* by Esbern Friis-Hansen and Charles Aben, is concerned with how adaptation to climate change is articulated politically at the subnational level, in particular by local governments. A discussion in 2013 among politicians from nine districts affected by floods in eastern Uganda revealed that climate change was generally not discussed at the district council level and was perceived as a 'lose–lose issue', that is, an issue that does not easily win votes and one with no central government financial resources attached to it. There was one exception, however, among these nine districts: Amuria District, where a politician had been elected as District Council chairman on a climate change adaptation platform. The chapter seeks to understand how it was possible to put climate change on to the political agenda of local government in this case. It goes on to explore how political 'will' after the election, despite a lack of additional fiscal resources, was transformed into the enacting district environmental ordinance and the formulation and implementation of wetlands management plans. The chapter further explores how political will within local government can translate into socially inclusive and institutionally sustainable climate change adaptation. The study concludes that local government political support is important in stimulating the emergence of new local institutions capable of enforcing regulations that restrict harmful natural resource management by the local elite.

The third section focuses on political ecology and climate change adaptation at the subnational level in a context in which the rural population is affected by climate change hazards, but where at the same time state institutions are weak in terms of both human and fiscal resources. Local government frontline bureaucrats are among many actors and not necessarily the strongest. With an emphasis on institutional dynamics, the political ecology of climate change is explored through three chapters based on empirical studies from Teso Region in Uganda. Local formal watershed governance is constrained by inadequate human resources and insufficient national budgetary support under the decentralized local governance framework. However, local groups organized around the use of natural resources exercise power in a variety of arenas and at multiple scales, thereby influencing the implementation of regulations for the protection of natural resources.

Chapter 8, *Political Ecology of Climate Change Management in Rural Uganda* by Charles Aben, discusses the concept of political ecology in climate change management in the context of decentralized natural resource governance in rural Uganda. The presentation of the concept draws a connection between political, economic and social drivers and alterations to traditional access rights, livelihood systems and social relations within the Awoja watershed. The study draws on the views of other scholars on climate change management and brings together an array of views and contexts for the clear understanding of the concept of political ecology.

Chapter 9, *Local Political Processes and the Management of the Awojo Watershed in Eastern Africa* by Charles Aben, Esbern Friis-Hansen, Jacob Godfrey Agea and J.J. Okiror, explores the political character of natural resource management in a changing climate and, in particular, the local political process that influences the management of natural resources in the Awoja watershed in eastern Uganda. Using a cross-sectional survey design that involves both qualitative and quantitative approaches, the study explores the changing relationship between the well-being of rural people living within the watershed, their dependence on natural resources and processes of environmental degradation. The study further questions how the inadequate human resource and national budgetary support for local government staff constrains and influences formal watershed governance at the lowest level of action. The chapter further explores how the weak presence of the state at the community level gives rise to locally autonomous natural resource management groups and how these groups exercise power at multiple scales with the aim of influencing the implementation of natural resource protection regulations.

Chapter 10, *Competing Interests over Natural Resources and Adaptation to Climate Change: the Case of Rice Cultivation in the Gweri Wetlands* by Charles Aben, Esbern Friis-Hansen, Bernard Bashaasha and J.J. Okiror, analyses the extent to which, and how, the combination of two trends are responsible for the increased tension between different land-use groups that has emerged over the past decade: climate change and the commercialization of rice cultivation. The chapter aims to examine the interactions of different institutions and enterprises and the consequences for climate change adaptation in the Awoja wetland system. By studying the contestation among local groups over ownership of the wetlands, the chapter shows how conflicts over wetlands resources are being exacerbated by fuzzy land property rights and weak local governance institutions. Examining what roles are played by business people, local government bureaucrats and the political and economic elites of local communities shows how elite capture of weak local public wetland management institutions has accelerated the degradation of the Awoja wetland system.

Section 4 explores two new current emerging themes for climate change governance. Common to both themes is that they are important phenomena that interact with climate change. Further, for both internal mobility and trade, the character of the interaction with climate change in Africa is empirically under-researched. Meanwhile the two themes receive very different attention at the international level. Migration is currently one of the hottest issues on the international agenda, although with a strong bias towards cross-border climate migration. In contrast, the relationship between liberalized markets and extreme climate change events such as droughts receives little or no international interest.

Chapter 11, *Social and Institutional Dynamics of Mobility as an Adaptation to Climate Change* by Charles Aben, Esbern Friis-Hansen and Isaac Nakendo, seeks to understand how the relationship between climate change and mobility has evolved during the past two decades, from an understanding of climate change as a direct cause of mobility to a more complex understanding that includes multiple factors that are unrelated to climate change. The chapter examines mobility as an adaptation response to climate change hazards and analyses how climate-induced changes to settlement patterns influence livelihoods, as well as socio-economic and political processes and outcomes. The chapter empirically explores the role of climate change on mobility in the cross-border areas between Teso and Karamoja regions. Specifically, the study seeks to identify and document the causes of mobility and the changing mobility patterns, as well as identifying those who become more mobile. Secondly, the study sets out to document and analyse the consequences of mobility: how new forms of social organization have emerged, how resources are shared between immigrants and locals, the conflicts that have arisen, and how they have been solved. Thirdly, the research seeks to determine to what extent mobility affects the environment and its effects on production and marketing systems. Finally, the research will assess the implications of mobility for service systems, culture and socio-economic benefits.

The final chapter before the conclusion is Chapter 12, *Hoarders as Saviours: the Performance of Regional Grain Traders During and After Extreme Drought* by Godfrey Suubi and Esbern Friis-Hansen. Although in theory the market liberalization of trade in agricultural grains contributes positively to climate change adaptation by lowering the hazards associated with drought, empirical observations in Soroti District during the 2012–2013 drought suggest that the absence of trade regulations increases these hazards. This chapter examines the relationship between trade and climate change, showing how the regulation of trade can be a tool for climate change adaptation. The purpose of this study is to bring to the forefront some critical issues related to food commodity hoarding that are related to extreme climatic change effects requiring the attention of local and meso-level institutions. The chapter focuses on the extent to which climate change hazards associated with drought are exacerbated by the behaviour of the private sector (traders).

References

Christoplos, I., Aben, C., Bashaasha, B., Hari Dhungana, H., Friis-Hansen, E., Funder, M., Huong, N.T.T., Khatri, D.B., Lindegaard, L., Mweemba, C.E. *et al.* (2016) *Governing Climate Change Adaptation in Africa and Asia: Actors, Politics and Local Institutional Change*. DIIS REPORT 2016:05. Danish Institute for International Studies, Copenhagen.

Conway, D. and Mustelin, J. (2014) Strategies for improving adaptation practice in developing countries. *Nature Climate Change* 4, 339–342.

Friis-Hansen, E. and Kyed, H.M. (2009) *Participation, Decentralization and Human Rights: A Review of Approaches for Strengthening Voice and Accountability in Local Governance*. World Bank, Washington DC.

Funder, M., Pain, A., Friis-Hansen, E., Christoplos, I. and Lindegaard, L.S. (2014) Can local governments save the planet? DIIS Comment. Available at: http://www.diis.dk/node/1452 (accessed 4 January 2017).

IPCC (2013) Climate Change 2013: *The Physical Science Basis. IPCC Working Group I Contribution to AR5*. Cambridge University Press, Cambridge, UK.

Lesnikowski, A.C., Ford, J.D., Berrang-Ford, L., Barrera, M. and Heymann, J. (2015) How are we adapting to climate change? A global assessment. *Mitigation and Adaptation Strategies for Global Change* 20, 277–293.

Nalau, J., Preston, B.L. and Maloney, M.C. (2015) Is adaptation a local responsibility? *Environmental Science & Policy* 48, 89–98.

Vogel, B. and Henstra, D. (2015) Studying local climate adaptation: a heuristic research framework for comparative policy analysis. *Global Environmental Change – Human and Policy Dimensions* 31, 110–120.

Yilmaz, S., Beris, Y. and Serrano-Berthet, R. (2008) *Local Government Discretion and Accountability: A Diagnostic Framework for Local Government*. World Bank, Washington DC.

Image courtesy of Tine Engedal

2 Implementing African National Climate Change Policies

Esbern Friis-Hansen

If economic marginalization is reinforced by the manner in which climate finance is utilized, and if social and cultural exclusion is not addressed by the manner in which access to public service, resources and assets is addressed through local government, then meeting the challenge of climate change would be continued "to be difficult".

(Webster, 2015, p.4)

From Climate Change Policy to Action

Climate change adaptation poses a new challenge for governance in Africa and is surrounded by considerable ambiguity. Policies addressing climate change adaptation have only recently emerged, driven by an international narrative, and are not widely understood or accepted by all stakeholders. This chapter aims to identify the drivers behind climate change policies, the intentions for their implementation, and the balance between national and subnational administrative levels with regard to control over finance, decision making and action.

The emphasis in international climate change fora during the past decade has been to agree on a joint policy framework for climate change mitigation and adaptation (UNFCCC(a) [no date]). The emphasis in and progress of these international narratives have had a strong influence on the evolution of climate change as a new policy area in Africa. This chapter first traces the links between the international climate change agenda and policy developments in Africa.

International discussion has increasingly turned to how support for climate change adaptation can be provided at the level of those who are vulnerable to climate change hazards (Conway and Mustelin, 2014; Funder *et al.* 2014; Lesnikowski *et al.*, 2015). This chapter examine the governance mechanisms associated with national climate change policies in Africa.

International researchers are increasingly concerned with the governance of multiple levels of climate change adaptation, particularly the role of subnational institutions (Nalau *et al.*, 2015; Vogel and Henstra, 2015; Christoplos *et al.* 2016). The chapter analyses rural climate change adaptation using the principle of subsidiarity.

State support for climate change action is likely to be ineffective if fiscal resources and decision making are disproportionally located within central administrations, which are not ideally placed to deliver an enabling political and technical environment in which rural communities can adapt to climate change hazards, nor do they have the capacity to do so. The chapter ends by assessing the balance between decision making and implementing power between national and subnational government institutions.

The International Climate Change Narrative and African National Climate Change Policies

During COP15 in Copenhagen in 2009, the first proposals were made for a global green climate fund (GCF) 'to support a paradigm shift in the global response to climate change' (GCF [no date]). The fund was established the following year at COP16 in Cancun with the commitment to collect US$100 billion by 2020 globally. This agreement is for the long-term financing of mitigation and adaptation to climate change with special attention given to Least Developed Countries (LDC), Small Island Developing States (SIDS) and African states, the most vulnerable countries. It is the outcome of very long, slow and politically conflict-ridden international negotiation. At the 21st COP held in Paris in 2015, the countries decided to draw up a new clear roadmap for reaching the US$ 100 billion by 2020 and, furthermore, decided upon setting a new goal after 2020 for future climate finance by 2025 (UNFCCC(b) [no date]).

COP 21 constitutes a climax after 20 years of multilateral climate change negotiations. A global climate change framework is discussed under the United Nations Framework Convention on Climate Change (UNFCCC) and the associated Conference of Parties (COP), whereas the scientific knowledge and narratives have been developed and agreed upon by the Intergovernmental Panel on Climate Change (IPCC) (UNFCCC(a) [no date]). In parallel to these multilateral negotiations setting global goals for CO_2 reduction and adaptation action, a local funding mechanism has been established to finance climate change action in developing countries. UN framework funds include the Green Climate Fund, the Least Developed Country Fund and the Adaptation Funds, whereas other funds such as the Local Environmental Fund are managed by the World Bank (UNFCCC(c) [no date]; World Bank [no date]).

An underlying implicit aim of governments and central ministries of the national climate change policy processes among African countries has been to become eligible for access to international climate finance from GCF. Therefore complying with the eligibility criteria for accessing international climate finance has strongly influenced the timing, format and content of climate change policies. The influence of international climate change narrative over the timing, format and content of national policy has been further strengthened by the manner in which policy formulation has taken place. Not unlike many other processes of policy formulation, it has been sponsored by international donor agencies, guided by a committee with representatives from relevant central ministries along with donor representatives under the auspices of the relevant ministry. Much of the actual drafting of policy background documents has been done by external consultants sponsored by development partners.

The policy formulation process in Zambia exemplifies this. A recent study finds that:

> Donors have played a central role in nurturing and influencing the climate change agenda in Zambia, and in developing the institutional framework for disaster management and climate change. However, the climate change agenda in Zambia is not only a donor construction: Although the issue is not high on the political agenda, there are de facto government interests related to climate change. This includes securing funding for civil service activities on the ground; addressing sensitive political aspects of disasters and food security; and dealing with possible threats to the national economy and political stability.
>
> (Funder *et al.*, 2013, p.4)

Closely synchronized with progress in the international COP negotiations, three waves of climate change policies have emerged simultaneously: National Action Plans for Adaptation (NAPA); National Climate Change Policy (NCCP); and Intended National Determined Contributions (INDCs) (Funder, 2016).

NAPAs were prepared by nine LDCs throughout Africa from 2006 to 2007 in response to UNFCCC (Agrawal and Perrin, 2008) with the overall aim of accessing financial support from the Least Developed Country Fund (LDCF). NAPA was at the time seen as a process whereby LDCs could identify and formulate a pool of project proposals that addressed their immediate climate change adaptation needs and vulnerabilities and for which national governments could attract international climate funding (UNFCCC(d) [no date]). A review of the project portfolio of 18 NAPAs (half from African countries) concluded that 'most of the projects in the NAPA documents seem far more aimed at building the capacity of national governments and agencies to coordinate adaptation to provide services to the general population, or to create infrastructure rather than to strengthen the capacity of local actors and institutions to undertake adaptation' (Agrawal and Perrin, 2008, p.13).

For example, the Uganda NAPA was prepared in 2007 by the Department of Meteorology in the Ministry of Water and Environment (MWE) in its capacity as the National Climate Change Focal (NCCF) point under the convention, with financial and consultancy support from GEF and UNEP. The foreword states that 'preparation of the Uganda NAPA was guided by the principle of participatory approach, drawing heavily on the views of the vulnerable communities and their knowledge on coping mechanism' (UNFCCC, 2007). Closer scrutiny of the Uganda NAPA document reveals, however, that in practice the participatory approach was limited to a consultative participatory rural appraisal (PRA) exercise among people in the areas visited by the formulation consultant team. No systematic inclusive planning process involving subnational government institutions was carried out. Local government staff and political leaders from nine districts in Teso and Karamoja regions who were interviewed in 2011–2013 confirmed that they were not consulted and had no influence over the formulation of the NAPA (Friis-Hansen *et al.*, 2013). The centralized formulation process of the Uganda NAPA also reflected its recommendations of nine technical priority areas proposed by central ministries to a total value of US$40 million. The Uganda NAPA was never implemented, however, because no finance was made available from LDCF as a result of continued disagreements at the COP negotiations (Malikut, 2014).

In this study we have identified and accessed online 11 African National Climate Change Policies (NCCPs) that were developed between 2010 and 2013. The overall stated purpose of NCCPs is to ensure that all national stakeholders address climate change impacts and their causes through appropriate measures, while promoting sustainable development and a green economy. Like the NAPA, the formulation of an NCCP has the underlying aim of making countries eligible to receive international finance from the newly established GCF. A review of the 11 NCCPs accessible online[1] indicates that they require a much more comprehensive analysis than a NAPA. However, although the planning process was somehow more inclusive, it does not include subnational government structures.

The Uganda National Climate Change Policy was formulated in 2012 as part of a process driven by a newly established Climate Change Unit (now a fully fledged department) in the Ministry of Water and Environment (MWE) and financially supported by international development agencies, including the EU, Department for International Development (DfID), Danida, the World Bank and the World Food Program (WFP). The draft policy document was produced by a group of national consultants guided by a committee including representatives from the central ministries, along with Danida representing the donor group. A number of consultative meetings were held by the consultancy team, who made efforts to include the perceptions of stakeholders at all administrative levels (Consultants, 2013, personal communication). Nevertheless, the NCCP formulation process was regarded by local government staff and politicians as essentially a top-down process because they were ignored during formulation of the implementation framework and therefore had no voice in determining the level of devolution of finance and decision-making authority (Local government and politicians in Teso, 2013, personal communication). The result is an NCCP that is not compliant with the decentralization act and that largely reduces the role of elected district councils and local government administrations to implementing agents with little or no influence or ownership.

All countries globally were asked to submit Intended National Determined Contributions (IDNCs) at COP 21 in Paris in 2015. An analysis of nine African IDNCs for which estimates of costs are available indicates that the full implementation of mitigation and adaptation priorities in NCCP is made conditional on the financial support of the international community (Table 2.1).

It is interesting to note that the national cost estimates in the IDNC are many magnitudes higher than the costs estimate in the

Table 2.1. Expected cost of climate change action in selected African countries until 2030 (billion US dollars).

	Ethiopia	Ghana	Kenya	Mali	Namibia	Niger	Rwanda	Uganda	Zambia
Total adaptation		12.8	7.5	1.0	19.6	1.6		2.4	15.0
Domestic		4.2				0.3			
International		8.6				1.3			
Total mitigation		9.8	5.3	34.7	10.4	7.0		5.4[a]	35.0
Domestic		2.0				0.8			
International		7.8				6.2			
Total	150.0	22.6	12.8	35.7	33.0	8.7	24.2	7.6	50.0

Source: Figures compiled from Intended National Determined Contribution reports accessed from UNFCCC (2015a).
[a]Need estimate for 2025 only.

NAPA less than a decade earlier. For example, the estimated cost of adaptation in Uganda is 72 times higher in the IDNC as compared with the NAPA. A smaller part of this difference can be explained by the period covered by IDNC, which at 15 years is three to four times longer than in the NAPA. The dramatic increase in the estimated cost of adaptation may also be explained by an increased level of understanding of the issues involved among central ministries staff and consultants based on the comprehensive analysis in NCCP (as far as we know, the global estimates of cost for mitigation and adaptation are also continuing to increase). Finally, the estimates may be inflated by the possible availability of the US$100 billion promised to be made available to the GCF by 2020. The cost estimates in IDNCs were calculated without countrywide consultation processes by central ministries and do not necessarily represent either an accurate or the most cost-effective approach to climate change adaptation (Table 2.2; Friis-Hansen *et al.*, 2015b).

The Zambian IDNC states that:

> It is estimated that over USD 50 billion is required for both mitigation (USD 35 billion for Domestic efforts with substantial International support) and adaptation (USD 20 billion) action across the programs up to 2030. Of this, USD 15 billion with the unconditional support provided by the Zambian government and USD 35 billion would be conditional support to be sourced externally. Something that would require international support in the form of finance, investment, technology development and transfer, and capacity building to fully realize its intended contribution.
>
> (UNFCCC, 2015b)

Table 2.2. Additional implementation costs for climate change adaptation in Uganda.

Sector (central ministry)	Million US$
Agriculture	297
Water	203
Fisheries	183
Transport and works	1054
Forestry	24
Wetlands	2
Biodiversity	6
Health	733
Energy	382
Wildlife and tourism	24
Human settlement	13
Disaster risk management	12
Vulnerable groups	4
Total	2919

Source: MWE, 2015.

Governance Mechanisms for National Climate Change Policies

Who controls access to financial resources and who takes decisions over the content of climate change action is therefore not only a question of how effective the implementation process is, but also of what type of action is given priority and ultimately whether supported activities make a difference to those affected by climate change hazards.

Review of the 11 African NCCPs available online show that limited attention is given to governance, which is usually described in vague language. In the 170-page-long NCCP for Tanzania, only one paragraph describes governance mechanisms: 'The general implementation of the strategies at sectorial level will be the responsibility of the relevant Government Departments and Agencies. The Prime Minister's Office – Regional Administration and Local Government (PMORALG) will work closely with Local Government Authorities (LGAs) through their various departments in collaboration with lined sectorial ministries to implement the strategic interventions at local level. The established cross-sectorial committees, which are the NCCSC and the NCCTC, will be important fora to facilitate implementation of cross-sectorial climate change interventions.' The use of vague language in this description of a governance mechanism makes it difficult to interpret who controls the financial resources and who decides what action to take.

Common to all 11 African NCCPs is that they ignore the decentralization reforms that have dominated the continent for the past two decades. None of the NCCPs explicitly specify the extent to which decision making and financial resources should be devolved to subnational governance institutions, but keep the description of governance mechanisms conveniently vague.

This can be illustrated by Uganda, where decentralization has been a dominant reform trend for the past two decades and where support for climate change would, if compliant with the Decentralization Act, fall within the mandate of local government. The Uganda National Climate Change Policy has coordination mechanisms at both the national and local government levels. At the national level, the climate change department in MWE has a coordinating function, while other central ministries are represented in a central coordinating committee and will be accountable for the implementation of the prescribed policy responses that concern them. At local government level, the focal point is anchored within the Natural Resources Department that is tasked with ensuring that climate change issues within all departments are integrated into District Development Plans. The existing District Environment Committee is expected to act as a mechanism to ensure cross-sectorial coordination (MWE, 2015).

Uganda's NCCP is, however, vague about the role intended to be played, if any, by the administrative leader of local government (Chief Administrative Officer) and the elected political leaders of local government (district councillors). The likely effect of the Uganda NCCP is to disempower the administrative and political leaders of local government and pursue a system of de-concentration rather than devolution, whereby the central ministry, in the Ugandan case the MWE, retains control of financial resources and uses environmental staff at the local government level as implementing agents (Friis-Hansen *et al.*, 2015a).

Such a model of de-concentration is not likely to be an effective mechanism for implementation. The empowering of the District Environmental Committee to coordinate and mainstream climate change activities for all other local government departments reflects the fact that the parent ministry, the MWE, formulated the policy. It will not, however, be easily implemented, as the district environmental officer is employed by and reports to the district chief administrative officer and not to the central ministries. The likely implications of a recentralization governance model of this sort are discussed under the heading The Disconnect between National and Local Levels of Governance with Regard to Fiscal Resources, Decision Making and Action.

The following four sets of possible explanations may shed light on why climate change decision making and access to finance are de-facto centralized:

1. Exclusiveness of policy formulation process. Centralization can be understood as an effect of the absence of downward accountability during the policy formulation process. The popular participation approaches used by the NCCP continue to be widely used in international development cooperation, but they have been subject to considerable criticism during the past 15 years. More often than not this approach to participation is reduced to a set of procedures in the policy formulation process, reflecting a technical rather than a political approach, with the empowerment of the beneficiaries being reduced to a consultative process within a set menu of choices. Hickey and Mohan's (2004) book *Participation as Tyranny* first synthesized the growing criticism of popular participation, arguing that such an approach is framed narrowly as a methodology to improve the targeting of the beneficiaries of a project or policy, rather than a process of fostering critical consciousness and decision making as the basis for active citizenship.

If participatory planning were to be taken seriously as a means of fostering empowerment, it would be implemented as a mutual decision-making process, with different actors sharing power and setting agendas jointly according to common goals (e.g. cost-effectiveness, subsidiarity). This discussion will be taken further in the next section.

2. Control over access to finance. After 20 years of decentralization reform that has purposely disempowered the central ministry's control over access to financial resources, climate change has become a new contested policy area through which power can be reversed to the national level. Central ministry actors are engaging with climate change as a technically complex field in which expertise exists primarily at the central ministry level and within national universities and research institutes. The logical effect of this understanding is that the bulk of capacity development should take place at the national level. This contestation has been further fuelled by the pledges made at international COP negotiations for dramatic increases in financial support. For many African countries, the potential for increased climate finance needs to be understood in the context of falling overseas development assistance. In Uganda, overseas development assistance has accounted for a large but declining share of the country's budget. As a percentage of the total budget, it declined from 52% in the fiscal year of 2003–4 to 30% in 2008–9 and is now less than 20%. In this context, international climate finance is highly attractive for central ministries as a replacement for overseas development assistance.

3. Central government path dependency. Part of the explanation for why government cabinets have approved NCCPs that centralize decision-making power and control over access to financial resources might relate to the governance mechanisms of the related policy area of emergency response. Emergency response is highly politically sensitive because the social contract between government and people dictates that the former take decisive action when the latter are in distress, for example, following natural disasters such as extreme floods and droughts (Christoplos *et al.*, 2014, 2016). Because of its politically sensitive nature, emergency response is typically highly centralized and placed directly in the Office of Prime Minister (OPM) or President's office. With the emergence of climate change adaptation as a policy field during the past decade, national politicians and central ministries have tended to equate climate change adaptation with emergency assistance and to replicate this sector's institutional roles and responsibilities at the national, meso and local levels.

4. Ambivalent international engagement with decentralization in support of climate change adaptation. Decentralization refers to different levels of the devolution of authority, responsibility and finance for public functions from the central government to intermediate and local governments at subnational administrative levels. Most African countries have undertaken decentralization reforms over the past two decades. The nature of reforms and approaches varies greatly, ranging from simple technical adjustments to public administrations in the form of de-concentration to radical distributions of political power between central and relatively autonomous local governments, referred to as devolution. International development agencies, in particular the World Bank, have invested heavily in support for devolution in order to counter bad governance practices rooted in corruption, authoritarianism and state bureaucracies. However, often central governments are not fully implementing their own plans for devolution (Crook, 2003; Ribot, 2007). Nevertheless, many local governments today provide a range of services that influence the quality of people's lives, including primary and sometimes secondary health care, primary and often secondary education, local infrastructure and agricultural advisory services (Friis-Hansen and Kyed, 2009).

The COP21 agreement emphasizes that central governments cannot and should not be the only movers on climate change. A range of other actors, including cities, local governments, civil society and the private sector, are key to making both mitigation and adaptation happen in terms of actual implementation, maintaining the pressure on national governments to address climate change generally (Christoplos *et al.*, 2016).

However, aside from references to subnational government in the COP 21 document and the 2014 IPCC report, international climate change support has until now ignored the rural subnational governance level. This can be seen from the lack of engagement with decentralization during policy formulation, which development partners have had considerable opportunity to influence. It can also be seen from the emphasis on the project mode of investments favoured thus far by development partners and the GCF. The project mode of implementation is well suited to the national centralized government approach associated with NCCPs in Africa. By using implementing through projects, central government ministries retain control over finance and decision making even when actual activities largely take place at the subnational level or local level. Line ministries also forge closer relations to their sister departments at local government level through project implementation, e.g. the MWE and Department of Environment at local government. The project mode also seems to be preferred by UN agencies, which generate considerably higher overheads from this approach compared with a devolution model that allows local government discretionary access to climate finance.

Chapter 5 of this book analyses the MWE and UNDP's project in Mbale, Uganda, where the project mode of implementation has resulted in the creation of parallel structures at the local government level and has bypassed the local government political context.

Subsidiarity Analysis of Support for Rural Climate Change Adaptation in Africa

The previous section determined that governance mechanisms in NCCPs are highly centralized and outlined four likely explanations for this. However, these explanations take their point of departure in contestations of political power and not in an analysis of how government support for climate change adaptation for rural citizens can be best delivered in a way that is cost-effective and socially inclusive.

The principle of subsidiarity states that the functions of government should be allocated to the lowest levels that are most competent to handle them (Ndegwa, 2002). On the one hand, for the case of climate change adaptation, the appropriate level of governance is necessarily local because the vulnerabilities associated with climate change hazards are highly context and location specific, and the process of adapting highly discretionary in nature and moreover likely to vary over time. On the other hand, the level of governance cannot be too local. Experience of autonomous adaptation shows that, while farmers have a long tradition of adapting natural resource management practices to changing weather conditions, they are unable to cope with the magnitude of changes associated with gradual climate change and extreme climate change events without collective action and assistance from administrative levels above the community.

A subsidiarity analysis seeks to determine which administrative levels of government and non-governmental actors can best deliver the required enabling environment that allows people to adapt to climate change. In other words, it asks what types of services, resources and assets are provided by which levels of government to ensure that both poor and vulnerable households and rural communities in general are served most effectively and efficiently.

The purpose of subsidiarity in the governance of climate change adaptation is to ensure that the most competent level of government is engaged in delivering the right type of services, resources and assets. Subsidiarity in the governance of climate change adaptation identifies the right level of government that is able to engage citizens in planning, budgeting, monitoring and evaluating, and implementing climate change action. Downward rather than upward accountability, and in particular social accountability to rural people, is an essential element of effective government support to climate change adaptation (Friis-Hansen and Ravnkilde, 2013).

The purpose of devolution of decision making and control over resources to the lowest appropriate level is to ensure that each level of governance engages in roles that are mutually beneficial to public employees at all levels, as well as to citizens. This way, each level of governance, whether it is a central ministry, local government or subdistrict government, contributes with specific technical, social and political capacities to the governance of climate change adaptation.

For rural households, climate change is a question of adapting to anticipated hazards that they may or may not experience in the coming season. To adapt, households must increase expenditure today to reduce the hazard that may or may not happen in the future. Better-off farmers are better able to cope on an individual basis, whereas resource-poor farmers are likely to have lower household adaptive capacity. It can therefore be argued that the provision of an enabling environment that allows rural households to address climate change hazards individually and collectively is an effective approach for government institutions to adopt in supporting climate change adaptation. As climate change hazards are locality specific and highly discretionary, effective climate change adaptation must ensure the interests and active participation of the local community, including the poor, as part of the participatory planning process (Friis-Hansen and Kyed, 2009). The empirical studies included in this book show that subnational political and administrative institutions, in particular at the district and ward/sub-county levels, have the administrative capacity and political legitimacy to create an enabling environment for climate change adaptation. Chapter 7 in this book provides an example of a local government that created an invited political space at subdistrict level that resulted in the generation of new content of existing institutions and the emergence of new institutions through a process of institutional bricolage (Cleaver, 2012).

It is important to examine the role of institutions, especially local ones, in shaping climate change adaptation if adaptation to climate change is to help the most vulnerable social groups. The effectiveness of adaptation to highly discretionary climate change hazards depends on the structuring of incentives for individual and collective action through local and extra-local institutions. Not only did existing institutions affect how rural residents responded to environmental challenges in the past, they are also the fundamental mediating mechanisms that will translate the impact of external interventions to facilitate adaptation to climate change. Institutions influence the local context-specific effects of climate hazards on livelihoods by structuring environmental risks and variability, and by creating an incentive framework within which the outcomes of individual and collective action unfold. Institutions are the media through which external interventions are reinforced (Webster, 2015).

Participation in decision-making processes linked to local government is seen by many as a means to secure more accountable governance and thus regarded as a critical development. Participatory planning and development is better able to provide relevant policies and local ownership of action, and therefore more likely to be effectively implemented, more likely to reach their intended outcomes and more likely to generate sustainable institutions to ensure long-term impact. Discussions of participation as a conceptual framework for development include downward social accountability, citizenship, social inclusion and rights (Hickey and Mohan, 2004; Gaventa, 2006; Friis-Hansen and Kyed, 2009). However, while participatory planning may provide opportunities for more sections of society to become involved in decision-making processes, whether or not this results in positive development outcomes depends on a range of other factors, including political support from local government.

Participatory planning of local government support is particularly relevant for services where the subject matter is highly discretionary (i.e. services that are not universal in character and are not easily standardized, such as agricultural extension services to small-scale farmers operating under diverse conditions or climate change adaptation) and intensely interactive (i.e. services that do not take the form of a one-off delivery but that require frequent interaction between the service provider and active citizens) (Pritchett and Woolcock, 2004). Adaptation by rural people in Africa to increased variability in seasonal weather patterns and extreme events in the form of floods and drought is highly discretionary and intensely interactive. A participatory approach that involves rural citizens in identifying problems, planning and implementing actions is therefore highly relevant.

Participation mediates socio-economic inequalities between local government officials and participants, as well as among the participants themselves. Many other community inequalities influence participatory planning, however, including gender, age and position of membership within the clan or tribe. Such inequalities may shape the form in which citizens participate, as well as institutional practices and outcomes. It is well known from the literature that participatory planning is often implemented in an instrumental manner that reproduces existing inequalities and avoids addressing resource conflicts (Mosse, 2001; Gaventa, 2004).

To conclude, subsidiarity analysis is about allocating different aspects of climate change adaptation at the level of governance most competent to handle it. Determining how to share financial resources between mitigation and adaptation is best placed at the global level: the Board of Green Climate Fund has decided on a 50:50 ratio between mitigation and adaptation (GCF, 2016). Developing national climate change policy, coordinating activities and providing technical backstopping are best undertaken at the central ministry level. Finally the appropriate level of governance for implementing rural climate change adaptation is necessarily local because the vulnerabilities caused by climate change hazards are highly context and location specific, and the process of adapting is highly discretionary in nature and likely to vary over time. On the other hand, the level of governance cannot be too local. Experience of autonomous adaptation (e.g. at the individual household level) shows that, while farmers have a long tradition of adapting natural resource management practices to changing weather conditions, they are unable to cope with the magnitude of changes associated with climate change without collective action and an outside enabling environment. The case studies in this book identify local government at the district and subdistrict administrative levels as the appropriate and most cost-effective level of decision making and action in the planning and implementation of climate change adaptation. However, the creation of an enabling environment for effective adaptation requires local government to create an adequate invited political space for community representatives to engage in a participatory planning process.

Climate change adaptation is intrinsically a political process because adaptation is closely linked to questions of power and natural resource management. In particular, for climate-change-induced emergencies, it is important that politicians at both the national and local government levels are seen to respond. Central ministries thus clearly have a role to play in setting the overall climate change policy framework and engaging in international fora such as COP negotiations. However, as institutions, central ministries are poorly placed to play a leading role in implementing climate change action because they have little or no relationship to or understanding of locally specific contexts.

The argument used by those in favour of the centralized governance of climate change adaptation is that only national ministries and research organizations, such as the National Meteorological Authority, have the technical capacity to determine the correct technical solutions and infrastructural investments to address climate change hazards. However, because of the variability of climate change hazards in space and time and among different social groups, the attempts to scale up standardized technical solutions are highly likely to fail.

The Disconnect between National and Local Levels of Governance with Regard to Fiscal Resources, Decision Making and Action

Access to and control of finance should be located at the level of government that is best suited to putting it to good use. Adhering to this principle of subsidiarity is crucial for cost-effective climate change adaptation. Currently, most African governments view financial resources for climate change as additional costs above ordinary development costs for each sector. African NCCPs identify incremental costs by incurred sector that will affect investment programmes and recurrent spending. Climate change policies identify the additional top-up requirement for each sector and largely expect this additional funding to come from external sources. This line of thought is based on the view that climate change adaptation can be solved by science and technology. Approaching adaptation to climate change through investments in infrastructure or technology and using a blueprint planning process and centralized implementation of projects is, however, likely to encounter resistance at the subnational level. In addition, in this blueprint model, rural citizens are reduced to being passive rather than active subjects.

The analysis of the 11 available NCCPs for Africa indicates that their more or less explicit implementation strategies represent a de-concentration approach that is likely to be noncompliant with their countries' decentralization legislation. When fiscal resources and political power are centralized at the ministerial level, the role of local governments as implementing agents, rather than as administrative and political partners, is reduced. Neither the de-concentration approach nor the UN project approach is cost-effective. Alternative mechanisms for channelling finance to the district and sub-county levels are needed. If district councils are bypassed by central ministries or UN projects, institutional sustainability is likely to be low. Climate change adaptation is highly discretionary and requires people's direct involvement in decision making and control of resources.

The centralization of decision making and control over funding by national climate change policies in central ministries creates a disconnect with those layers of government where climate change action happens. This centralization of power and resources undermines attempts to increase effectiveness by applying subsidiarity principles to the governance of climate change adaptation in a way that allows local governments the political space to become involved in participatory planning processes that mobilize local citizens. Centralized control of funding and decision making disempowers local people and limits the possibilities for participatory planning.

International debate (Ribot, 2014; Taylor, 2015) suggests that support for an enabling environment that allows rural people to adapt to climate change should engage with their livelihoods in ways that change the social economic relationships that shape and condition those livelihoods. Taylor (2015) argues that climate change should not be 'passed out and isolated from the ongoing processes of social and ecological transformation that constructs our lived environment' ((Friis-Hansen *et al.*, 2015c) Taylor, 2015, p.xii).

Matching the subsidiary analysis to our understanding of the character of 'climate' through political ecology suggests that the character of local governance is crucial in managing and utilizing the politics of climate change adaptation. Applying Taylor's understanding of 'climate' to the African context, local governments become the crucial unit at adaptation because climate can be understood as the production of sustainable natural resource management.

Note

[1] NCCPs are available online from Ethiopia, Malawi, Ghana, Kenya, Mali, Nigeria, Tanzania, Uganda, Zambia, Rwanda and Niger.

References

Agrawal, A. and Perrin, N. (2008) Climate adaptation, local institutions, and rural livelihoods. International Food Policy Research Institute. *Working Paper* W081-6, University of Michigan, Michigan.

Christoplos, I., Aben, C., Bashaasha, B., Dhungana, H., Friis-Hansen, E., Funder, M., Huong, N., Khatri, D., Lindegaard, L., Mweemba, C. *et al.* (2014) Towards "good enough" climate and disaster governance. Emerging lessons from Zambia, Nepal, Viet Nam and Uganda. *DIIS Report*, 2014:21. Danish Institute for International Studies, Copenhagen.

Christoplos, I., Aben, C., Bashaasha, B., Dhungana, H., Friis-Hansen, E., Funder, M., Huong, N., Khatri, D., Lindegaard, L., Mweemba, C. *et al.* (2016) Understanding sub-national climate governance. Findings from Nepal, Uganda, Vietnam and Zambia. *DIIS Report*, 2016:5. Danish Institute for International Studies, Copenhagen.

Cleaver, F. (2012) *Development through Bricolage. Rethinking Institutions for Natural Resource Management.* Earthscan, Routledge, Abingdon, UK.

Conway, D. and Mustelin, J. (2014) Strategies for improving adaptation practice in developing countries. *Nature Climate Change* 4, 339–342.

Crook, R.C. (2003) Decentralisation and poverty reduction in Africa: the politics of local–central relations. *Public Administration and Development* 23, 77–88.

Friis-Hansen, E. and Kyed, H.M. (2009) *Participation, Decentralization and Human Rights. A Review of Approaches for Strengthening Voice and Accountability in Local Governance.* World Bank, Washington DC.

Friis-Hansen, E. and Ravnkilde, S.M.C. (2013) Social accountability mechanisms and access to public service delivery in rural Africa. *DIIS Report,* 2013:31. Danish Institute for International Studies, Copenhagen.

Friis-Hansen, E., Bashaasha, B. and Aben, C. (2013) Decentralization and implementation of climate change policy in Uganda. *DIIS Working Paper,* 2013:17. Danish Institute for International Studies, Copenhagen.

Friis-Hansen, E., Aben, C., Bashaasha, B., Okiror, J.J. and Chelli, P. (2015) Local politics of climate change adaptation in Uganda. *Paper presented at 10th International Conference in Interpretive Policy Analysis*, 8–10 July 2015, Lille, France.

Friis-Hansen, E., Bashaasha, B., Aben, C. and Okiror, J.J. (2015) Local government engagement with climate change adaptation in Uganda. *DIIS Report,* 2015.19. Danish Institute for International Studies Copenhagen.

Friis-Hansen, E., Bashaasha, B., Okiror, J.J., Aben, C., Suubi, G., Rosset, S., Chelli, P. and Nakendo, P. (2015) Climate change adaptation lacks local government funding: Decentralized climate change responses in Uganda. *DIIS Policy Brief*, August. Danish Institute for International Studies, Copenhagen.

Funder, M. (2016) Policy-makers and climate financing in African LDCs: the case of Zambia. In: *Financing Sustainable Development: Actors, Interests, Politics*. Danish Institute for International Studies, Copenhagen.

Funder, M., Mweemba, C. and Nyambe, I. (2013) The climate change agenda in Zambia: national interests and the role of development cooperation. *DIIS Working Paper*, 2013:13. Danish Institute for International Studies, Copenhagen.

Funder, M., Pain, A., Friis-Hansen, E., Christoplos, I. and Lindegaard, L.S. (2014) Can local governments save the planet? *DIIS Comment*. Danish Institute for International Studies, Copenhagen. Available at: http://www.diis.dk/node/1452 (accessed 4 January 2017).

Gaventa, J. (2004) Strengthening participatory approaches to local governance: learning the lessons from abroad. *National Cicic Review* 93, 16–27.

Gaventa, J. (2006) Finding the spaces for change: a power analysis. *IDS Bulletin* 37, 23–33.

GCF (Green Climate Fund) [No date] The Big picture. Available at: http://www.greenclimate.fund/the-fund/the-big-picture#history (accessed 1 September 2016).

GCF (Green Climate Fund) (2016) Policies for resource allocation. Available at https://www.greenclimate.fund/documents/20182/239759/3.4_-_Allocation_of_Fund_Resources.pdf/e717fcd0-3f85-4a5d-81aa-9199c5df4c3d (accessed 13 September 2016).

Hickey, S. and Mohan, G. (eds) (2004) *Participation: From Tyranny to Transformation? Exploring New Approaches to Participation in Development*. Palgrave, New York.

Lesnikowski, A.C., Ford, J.D., Berrang-Ford, L., Barrera, M. and Heymann, J. (2015) How are we adapting to climate change? A global assessment. *Mitigation and Adaptation Strategies for Global Change* 20, 277–293.

Malikut, C. (2014) Status of climate change policy development, implementation, NAPA and NAMA in Uganda. http://www.slideshare.net/cgiarclimate/mc-presentation-ccafs29-april-2014 (accessed 4 January 2017).

Mosse, D. (2001) 'Peoples knowledge' participation and patronage: operations and representation in rural development. In: Cooke, B. and Kothari, U. (eds) *Participation: The New Tyranny?* Zed Books, London.

MWE (The Ministry of Water, Lands and Environment) (2015) Uganda National Climate Change Policy. Available at: http://ccd.go.ug/index.php/resources-publications/220-uganda-national-climate-change-policy (accessed 22 May 2016).

Nalau, J., Preston, B.L. and Maloney, M.C. (2015) Is adaptation a local responsibility? *Environmental Science & Policy* 48, 89–98.

Ndegwa, S.N. (2002) *Decentralization in Africa: A Stocktaking Survey*. World Bank, Washington DC. Available at: https://www.k4health.org/sites/default/files/Decentralization%20in%20Africa%20Report%20.doc (accessed 15 August 2016).

Pritchett, L. and Woolcock, M. (2004) Solutions when *the* solution is the problem: arraying the disarray in development. *World Development* 32(2), 191–212.

Ribot, J. (2007) Representation, Citizenship and the Public Domain in Democratic Decentralization. *Development* 50, 43–49.

Ribot, J. (2014) Cause and response: vulnerability and climate in the Anthropocene. *Journal of Peasant Studies* 41, 667–705.

Taylor, M. (2015) *The Political Ecology of Climate Change Adaptation*. Routledge, New York.

UNFCCC (2007) Uganda National Adaptation Programmes of Action. Available at: http://unfccc.int/resource/docs/napa/uga01.pdf (Accessed 15 May 2016).

UNFCCC (2015a) INDCs as communicated by parties. Available at: http://www4.unfccc.int/submissions/indc/Submission%20Pages/submissions.aspx (accessed 13 May 2016).

UNFCCC (2015b) Zambia's Intended Nationally Determined Contribution (INDC) to the 2015 Agreement on Climate Change. Available at: http://www4.unfccc.int/submissions/INDC/Published%20Documents/Zambia/1/FINAL+ZAMBIA'S+INDC_1.pdf (accessed 15 May 2016).

UNFCCC(a) [No date] Intergovernmental Panel on Climate Change (IPCC). Available at: http://unfccc.int/bodies/body/6444.php (accessed 31 August 2016).

UNFCCC(b) [No date] Newsroom: Historic Paris Agreement on Climate Change. Available at: http://newsroom.unfccc.int/unfccc-newsroom/finale-cop21/ (Accessed 1 September 2016).

UNFCCC(c) [No date]. FOCUS: Climate Finance. Available at: http://unfccc.int/focus/climate_finance/items/7001.php (accessed 31 August 2016).

UNFCCC(d) [No date] Background information on the NAPAs. Available at: http://unfccc.int/adaptation/workstreams/national_adaptation_programmes_of_action/items/7572.php (accessed 31 August 2016).

Vogel, B. and Henstra, D. (2015) Studying local climate adaptation: A heuristic research framework for comparative policy analysis. *Global Environmental Change – Human and Policy Dimensions* 31, 110–120.

Webster, N. (2015) *Paper presented at 10th International Conference in Interpretive Policy Analysis*, 8–10 July 2015, Lille, France.

World Bank [No date] Climate Finance – Overview. Available at: http://www.worldbank.org/en/topic/climatefinance/overview (accessed 31 August 2016).

BARUA PEPE YA
HALMASHAURI YA
WILAYA YA WANGINGOMBE
NI
ded.Wangingombe@
njombe.go.tz
ISS/TAB/No
2012

3 Contemplating Climate Change at Local Government: On-the-ground Politics of Adaptation Delivery in Tanzania

Sarah Ann Lise D'haen and Jonas Østergaard Nielsen

Introduction

With the recent proliferation of human and financial losses owing to extreme weather events and the increased awareness of the likelihood that major changes in the earth's system are already underway, adaptation rockets up the international policy agenda. As livelihoods and resources in the world's least developed nations are progressively more affected (IPCC, 2014), many of these countries are gearing up to identify priority areas and sectors to be targeted under various (global) financial mechanisms (Berrang-Ford *et al.*, 2014; Heinrich Böll Stiftung and ODI, 2015). After initial policy formulations such as the National Adaptation Programmes of Action (NAPAs), developing countries are now drafting more comprehensive national climate change adaptation policies and strategies. Amidst this heightened activity, questions arise about the practical delivery of adaptation policy on the ground (Conway and Mustelin, 2014; Lesnikowski *et al.*, 2015). In recent years, the scientific community has explored what 'actual' adaptation implementation might look like, how it could come about, why and how it might be challenged and how challenges could be mediated, overcome or avoided altogether (Berrang-Ford *et al.*, 2011). Increasing attention is being paid to national operationalization of multi-level adaptation governance systems (Adger *et al.*, 2005; Amundsen *et al.*, 2010) and with this the roles and responsibilities of different levels of government, and particularly those of local level government (Nalau *et al.*, 2015; Vogel and Henstra, 2015).

National policies are typically 'translated' into practical measures at local government and ample documentation exists of the barriers to mainstreaming adaptation at this level (Amundsen *et al.*, 2010; Pasquini *et al.*, 2013). However, wider issues are at play. In many developing countries a pervasive disconnection exists between national discourses and framings of adaptation and local realities (Conway and Mustelin, 2014; Sherman and Ford, 2014). Moreover, in most developing countries strategies are still awaiting translation into actual action plans and huge uncertainty exists as to how implementation will look in practice. Finally, widespread expectations for local governments to play a key role in adaptation delivery stand in stark contrast to the general absence of clear guidelines on their roles and responsibilities both in developed (Tompkins and Adger, 2005; Bulkeley and Kern, 2006; Crabbe and Robin, 2006; Urwin and Jordan, 2008; Wagner and Zeckhauser, 2012; Nalau *et al.*, 2015) and in developing countries (Agrawal *et al.*, 2013; Nam *et al.*, 2015).

The uncertain operationalization plans and the unclear adaptation mandate leave room for struggles over authority and legitimacy (Paterson *et al.*, 2003; Betsill and Bulkeley, 2006) and are thought to play out differently in developing and developed countries. In developed countries, where local governments are typically downwardly accountable, taking on adaptation responsibilities 'represents financial and political risks' (Nalau *et al.*, 2015) and involves ethical and moral considerations about spending limited public resources (Cashmore and Wejs, 2014). In Australia, for example, these issues lead local governments to shift responsibility to other tiers of government and to express a lower willingness to implement adaptation measures (Nalau *et al.*, 2015). In developing countries, however, local governments are typically upwardly accountable to line ministries, and depend on national earmarked allocations for their budgets (Ribot *et al.*, 2010). In the context of promising (global) financial mechanisms for funding adaptation, local government actors are thought to compete for adaptation responsibilities hoping to benefit from the financial resources and the increased political influence that might come with it (Conway and Schipper, 2011).

In this context, questions arise about how adaptation delivery is perceived and 'done' by local government actors in sub-Saharan Africa. What role is there for local government in adaptation and who at local government wants to work with adaptation and why? How do individual bureaucrats see their districts', their departments' and their own role in delivering adaptation?

To address these questions, this chapter explores the nascent operationalization of Tanzania's National Climate Change Strategy (CCS). It traces discourses of adaptation across, within and between different departments and tiers of the Tanzanian government and pays special attention to the framing of tasks and responsibilities in delivering adaptation. Combining interview data of civil servants' discourses on climate impact issues, the chapter aims to find out whether and which particular discourses can be linked to particular actors, their understandings, interests and (struggles for) power or influence. Ultimately, it is investigating how these results can inform the effective operationalization of adaptation implementation strategies in the Tanzanian context.

This introduction is followed by a brief overview of academic work on the politics of adaptation and local government. We then briefly set the scene before describing the data collection and research methodology. The Results section focuses on the role of local government in national adaptation action plans and on how local government bureaucrats navigate and contemplate (the expansion of) their mandate in the context of the national adaptation agenda. The final section reflects critically on these political processes of adaptation at local government in Tanzania and relates them to wider trends, processes and implications of adaptation politics in developing countries.

The Politics of Adaptation Implementation

The concept of adaptation and its imperative prescriptions are under close scrutiny. The portrayal of climate change as an urgent and serious external threat to society has facilitated the emergence of a climate industry characterized by increasing technicalization and professionalization processes (Arnall *et al.*, 2014) and actually allowing 'climate change to become governable as a managed process of adaptation' (Taylor, 2015). Responding to climate change is, however, closely related to politically and historically contextualized decision making in key environmental issues (Naess *et al.*, 2015) and thus calls for analysis of who claims adaptation and why. As climate change and adaptation are increasingly understood as discursively situated (Hulme, 2008), scholars from various fields have begun to uncover processes of decision making in policy implementation (Eisenack *et al.*, 2015), tracing framings and constructions of climate risks and adaptation by practitioners to unveil different understandings, interests and power issues (Aldunce *et al.*, 2015; Artur and Hilhorst, 2012). A central finding has been that actors engaged in various aspects of climate change adaptation react as much to the idea of adaptation as to actual impacts of adverse weather conditions on the ground (Weisser *et al.*, 2014). In exploring adaptation as a socio-political and cultural process embedded in historical and institutional contexts of societal transformation, this literature proposes concepts of knowledge, power and subjectivity (Granderson, 2014; Eriksen *et al.*, 2015; Naess *et al.*, 2015) for theorizing and conceptualizing how 'different actors are able to further their particular interests in adaptation actions, and how adaptation actions may both reinforce unequal power relations but in other cases open up spaces for contesting inequalities' (Eriksen *et al.*, 2015). In the context of historical underfunding of subnational institutions, departments and agencies related to the environment, the heightened international attention for climate change issues combined with the tendency of national governments to assign climate change responsibilities to the national minister for environment (Pasquini *et al.*, 2013) presents a new avenue through which local level environmental bureaucrats negotiate power with their colleagues in other departments and tiers of government. In this chapter, we take our cue from these insights and explore the importance of local politics at district level in shaping the adaptation implementation scene in Tanzania.

Setting the Scene

Tanzania's climate change policies and governance

Tanzania's legal and institutional basis for climate policy originates from the 1997 National Environmental Policy and the 2004 Environmental Management Act (EMA) (Division of Environment, 2012). Climate change policy is coordinated by the Division for Environment (DoE) in the Vice President's Office (VPO). Environmental departments and climate change desks in each line ministry oversee the integration of climate change in sectoral policies and plans and coordinate the development of Sector Resilience Plans that detail the practical implementation of the NCCS (Ministry for Agriculture, 2013). The DoE houses the National Climate Change Focal Point (NCCFP), and is assisted by the inter-ministerial National Climate Change Steering Committee (NCCSC) and the National Climate Change Technical Committee (NCCTC). It is further advised by the National Environmental Management Council, which has the power to monitor and enforce compliance with environmental standards (Greene, 2015). The Finance Ministry is responsible for the financial management (Division of Environment, 2012) through a National Climate Fund, expected to be filled by international donors (Division of Environment, 2012). Under decentralization, climate change at the local level officially falls under the responsibilities of the District Environmental Officer and the District Environmental Committee (Division of Environment, 2012). The integration and compatibility of climate activities and sectoral activities at district level is overseen both by the Prime Minister's Office (PMO), who is responsible for Local Government Authorities (LGAs), and by environmental officers at the Regional level Secretariat.

Three main climate-relevant policy documents have been formulated since the EMA: the 2007 National Adaptation Programme of Action (NAPA); the 2012 National Climate Change Strategy (NCCS); and the 2013 National REDD+ Strategy and Action Plan. Although these documents and the allocation of climate policy coordination responsibilities to the VPO reveal the general political importance adhered to climate issues in Tanzania, substantial uncertainty remains about operational implications in terms of identifying, financing and implementing adaptation needs and activities. The NAPA suffered from unclear priority identification, limited stakeholder involvement and, in general, a highly political agenda. The NCCS and REDD+ strategies have scored much better on these accounts but the science base for projecting subnational future climate impacts, as well as a national vulnerability assessment, on the basis of which priority areas or populations could be identified and against which progress could be measured, is still lacking . At the time of fieldwork, no separate climate finance management was in place in Tanzania. In fact no clear financing plan, identifying expected costs or potential sources of funding for specific actions, had been developed (Yanda *et al.*, 2013). The national level environmental desks are thought to lack the knowledge and financial resources to properly execute the NCCS's objective to coordinate and mainstream mitigation and adaptation activities across, but mainly within, existing and ongoing sectoral programmes, plans and strategies (Division of Environment, 2012); the NCCSC and NCCTC are found to lack political wielding power and moreover have failed to meet on a regular basis (Yanda *et al.*, 2013). Finally, the national Planning Commission, mandated to monitor, analyse and provide advice on long-term sector policies and socio-economic developmental issues at national level, is absent in the climate governance structure at national level (Yanda *et al.*, 2013).

The Great Ruaha river catchment

For the purpose of this study, we focus on local government actors within the Great Ruaha river catchment, located in south-western Tanzania (Fig.3.1). More specifically, we focus on the administrative bodies of two adjacent districts within the catchment – Kilolo and Iringa Rural District Council (Kilolo and Iringa District from here on) – and on the administration of the Rufiji Basin Water Office (RBWO). Kilolo District (population 220,000, residing in 106 villages) and Iringa District (population 250,000, residing in 123 villages) report to the Iringa Regional Secretariat, which has its headquarters in Iringa town. Originally part of Iringa District, Kilolo's district council was established in 2008 at new buildings in Kilolo, a then small highland village south of Iringa town. Iringa District kept the original headquarters in Iringa town. Finally, the RBWO also has its headquarters in Iringa town. This independent government agency is responsible for regulating water use in the Rufiji river basin to which the Great Ruaha catchment belongs.

The Great Ruaha and its tributaries are essential for the livelihoods and the local economy in the districts. Drinking water is almost exclusively sourced from these rivers and a considerable part of agricultural activities rely on irrigation schemes. The rivers furthermore support livestock production in the area. More

Fig. 3.1. Location of Kilolo and Iringa Rural district within the Great Ruaha river catchment (striped) and the Rufiji water basin in Tanzania.

than 90% of the population in the districts relies on agriculture for its income. Apart from growing subsistence food crops, households are engaged in animal husbandry and the production of horticultural products such as tomatoes and onions. Major commercial agricultural activities include irrigated rice schemes in Iringa District, wood lot production in Kilolo District and big dairy farms in both districts. The Great Ruaha also has national importance. Its wetlands form the basis of the ecosystems and wildlife of Ruaha National Park, and Mtera dam, located downstream in the east of Iringa district, is one of the country's major hydropower plants.

The climate in the study area ranges from cool and humid in the south-eastern highlands of Kilolo District to dry and hot in the northern, low-lying areas of either district. In recent years, increased intra-annual rainfall variability and a general increase in temperature combined with several floods events along the Likosi river in Kilolo District and recurrent dry periods and low water levels of the Great Ruaha in large parts of Iringa District have started to affect agricultural production. Apart from directly challenging subsistence food production, the general rise in temperature impacts growing seasons of highland commercial crops such as tree fruits, as well as pest occurrence in lowland commercial crops such as tomato and onion. Moreover, both floods and droughts are affecting alternative income sources such as tourism and have caused economic loss in the infrastructure and energy sectors, both locally and nationally. Future projections for Tanzania entail a further rise in temperatures, and an increase in rainfall especially in the wet season, with more frequent occurrence of heavy rainfall events (McSweeney *et al.*, 2010).

Data and Methods

This study is based on 22 semi-structured interviews with civil servants at district, ward and village level in Iringa and Kilolo District as well as with civil servants within the RBWO. The interviews were conducted in February and November 2013. We take an actor-centred approach and focus on rural bureaucrats' framing of job responsibilities in the context of the emerging adaptation agenda, relative to other responsibilities and other actors within and outside local government. For our analysis, we focus on district departments and government agencies whose mandate and activities touch upon natural resource management and environmental issues. To enhance understanding of broader issues and to triangulate information on practices and policies, four additional interviews were conducted with civil servants from Iringa Regional Secretariat, the national level Disaster Risk Reduction Directorate as well as with programme officers from the Tanzanian National Red Cross Society. The interview protocol was structured along five tiers asking: (i) how interviewees saw their everyday job, their responsibilities and challenges; (ii) how interviewees saw the relationship between the responsibilities of their own departments/sectors/agency and that of (other) district departments, including interpersonal relationships with staff from these departments; (iii) how interviewees saw the district level in the 'state'; (iv) what and how interviewees' thought about climate variability and change; and, finally, (v) what and how interviewees thought of adaptation, adaptation policies and the districts' role in planning, implementing or delivering these. Because one of the aims of the fieldwork was to find out if and how (district) civil servants considered climate change and adaptation to be of importance in (their) districts, climate change was never explicitly 'cued' during the first three tiers of the interview. All interviews were conducted in English, fully transcribed and subsequently coded and analysed in the qualitative data analysis software NVivo.

Results

Local actors in the Great Ruaha River Basin

Tanzania's districts officially have the authority to locally coordinate socio-economic development (Local Government and Reform Programme II; PMO-RALG, 2009). Next to departments directly representing and funded through core national policy domains such as agriculture, water, education and health, district governments include more locally anchored and funded departments for planning, community development (and social welfare) and natural resource management.

Home to the District Environmental Officer and coordinator of the District Environmental Committee, the district's Natural Resources Department (fully the 'Land, Forests and Natural Resources Department' at the time of fieldwork) is officially mandated to deal with climate change issues at local level. The department more generally oversees the management and protection of the districts' natural resources. Whereas the Land Section of this department occupies itself predominantly with (village) land use planning processes and the issuing of land occupancy certificates, the Forests and Natural Resources Section focuses on sustainable resource management and conservation issues. Officers describe their activities as aiming at initiating and facilitating alternative income sources for a population partially reliant on forest products. Projects in this context involve establishing small-scale tree nurseries for commercial tree growing and tree fruit production, as well as stimulating small-scale commercial beekeeping, honey production and mushroom cultivation (Interviews 5, 6, 11 and 12).

At the time of interview no official climate-relevant structures were in place at either Kilolo or Iringa District and very limited climate strategy had been communicated by the national government to the district level. Moreover, officers had no information on future climate projections and no comprehensive vulnerability assessments were available. None the less, from our interviews we derive that civil servants in the Natural Resources Departments are well aware of the recent climate policies such as the NCCS and the REDD+ strategy and were actively contemplating their mandate in dealing with mitigation and adaptation questions (Interviews 5, 6 and 12). Apart from the Natural Resources Department, several other district departments are relevant in this respect.

In a rural context, agriculture is one of the most dominant district departments both in terms of human as well as financial resources. The department's activities support the implementation of the District Agricultural Development Plan (DADP), a consolidation of village level development projects locally identified through a bottom-up participatory process called Opportunities and Obstacles for Development (O&OD) and repeated yearly in each of the district's villages. Co-funded either at farmer group or at community level, the O&OD projects in Iringa and Kilolo Districts aim to overcome technical barriers to (smallholder) agricultural production, either in terms of inputs – by building seed banks, establishing tree nurseries, procuring fertilizers, pesticides or herbicides – or in terms of physical infrastructure – by supporting the construction of irrigation schemes, rainwater harvesting dams, feeder roads or market buildings (Interviews 1, 2, 8, 10, 15, 17 and 20).

The district Planning Department is essential to the overall working of the district government. It is responsible for drawing up the districts' annual budget estimate and for the distribution of national funds to the individual development projects under the district's many sectoral programmes (Interviews 9, 15 and 16). Aiming to increase the district's control over its development path, this department additionally takes on a crucial role in developing and coordinating the district's domestic revenue-generating strategies (Interviews 15 and 16). In this context, it gathers and manages information about income-generating initiatives and commercial investment opportunities in land, forest and other natural-resources-related areas (Iringa District Council, 2012).

Whereas the planning department can be considered as the district's strategic coordinator, the practical enabling of the districts' projects is in the hands of the Community Development Department. Specialists in managing people, facilitating discussions and guiding target group formation processes, community development officers see themselves as indispensable for the successful implementation of the projects of, for example, the agricultural department (Interviews 1, 2, 8, 10 and 14).

A final player in the local adaptation debate is the RBWO, the official monitoring and regulating agency for water resources in the Great Ruaha River catchment. Directly reporting to the Minister of Water, the RBWO issues water use permits and manages a catchment water use calendar. Overall, the RBWO carefully balances a mandate to licence water use with one to control and protect water resources (Interview 23).

The way in which civil servants within these different administrations engage in the adaptation debate differs considerably. For example, whereas 'climate change' and 'adaptation' are frequently mentioned by civil servants in our interviews with the RBWO, and the district departments for Planning, Community Development and Natural Resources, these concepts are absent from our interviews with district agricultural bureaucrats (Interviews 1–6, 8–12, 14–16 and 23). The former bureaucrats frequently mention 'future', 'adaptation' and 'strategy' in connection to dealing with climate-related impacts (e.g. Interviews 5, 6, 8 and 16) in contrast to agricultural officers who connect relevant climate-related actions to following 'guidelines' and meeting 'priorities' of national agricultural policy (Interviews 1, 2, 10 and 17).

Against this background, the emerging climate change debate opens up an arena for contestation of power, struggles, manoeuvring, and positioning at local level. In the absence of a clear allocation of responsibilities and official climate-relevant management structures, the nascent climate change adaptation agenda is being used to justify certain actions or to justify non-cooperation between various actors in different departments and government agencies. In the next section we explore how local government actors navigate these uncertain spaces.

Sites of struggle: local bureaucrats contemplating dealing with climate impacts

One example of local struggles in the context of the adaptation debate can be found in the area of water resources. Coordination of water resources management in the Great Ruaha river catchment officially falls under the responsibility of the RBWO. With roughly ten officers based at its headquarters in Iringa town and responsible for a 183,791 square kilometre area (Ministry of Water and Irrigation, 2014), the RBWO is extremely understaffed, and is therefore assisted by village-level Water User Associations (WUA) who locally monitor water use by individual farmers and who resolve potential local water conflicts. The WUAs report back to the district, often the district agricultural department. With extension workers at ward and village level, agricultural departments have additional eyes and ears on the ground when it comes to water use by farmers. From our interviews we derive, however, that information-sharing between district agricultural officers and staff at the RBWO is virtually non-existent (e.g. Interviews 12 and 23). The authors' personal observations are of non-permitted reopenings of irrigation channels in an upstream village on the Great Ruaha, and of numerous unregistered portable irrigation pumps operating along the river at the time of fieldwork.

The cooperation challenges in monitoring actual river water use are mirrored in actors' diverging ideas about future water use in the basin. Agricultural officers in both districts have noticed the general rise in temperatures during the past decade and identify potential future impacts such as changes in growing seasons, more frequent food production shortages and increasing conflicts between farmers and livestock keepers. They consider dealing with these impacts as part of their mandate to 'assist farmers so they have enough food' (Interviews 1, 2 and 10) and suggest, amongst other technical solutions, to focus efforts on expanding existing and establishing additional irrigation schemes (Interviews 1, 2, 10 and 17). Conversely, the RBWO, relying on its mandate to 'protect future water resources', is sceptical about the physical potential for increasing irrigated agriculture under changing climatic conditions next to satisfying other demands such as domestic and industrial use, livestock production, wildlife ecosystem functioning and electricity production at Mtera dam (Interview 23). Even though they share their agricultural office's view that irrigation schemes are 'a good way to deal with dry areas and droughts' (Interview 15), the district departments for Planning, Community Development and Natural Resources altogether take a mixed stance on irrigation expansion (Interviews 5, 6, 8, 9, 11, 12 and 14–16). A recent 'trauma' in Iringa District made officers here acutely aware of the need to secure continued long-term water availability and carefully consider increasing multiple demands on river water. As part of its strategy to boost the local economy in the driest parts of the district, Iringa rural district built, from its own resources, a new fish market at the edge of the Mtera dam reservoir in 2009. Water levels in the reservoir started lowering the same year, however, and rendered the market building, located

2 km away from the edge of the lake at the time of visit in 2011, in effect useless. The halted fishing industry not only represents a substantial loss in income generation for local fishermen, but also for the district itself. Even though several district officers suspect excessive irrigation at commercial rice schemes upriver to be responsible for the lowering of the water level, the trauma has alerted them to the risk of increased river water use in general, also by smaller scale irrigation schemes (Interviews 15 and 16).

At the time of fieldwork, no platform existed where these diverging visions or uncertainties could be discussed. Moreover, as pointed out earlier in this chapter, no comprehensive climate projection information or vulnerability assessment was available. In fact, the lack of information sharing and cooperation at local level, but also across different levels and domains of government, was identified by individual bureaucrats in the RBWO and in the Community Development and Natural Resources departments (Interviews 12, 14 and 23): 'we are all government, but we are not together' (Interviews 12 and 23). Telling in this context is that in our interviews, Iringa agricultural officers mention that they foresee the need to negotiate with neighbouring Kilolo and Mufundi districts, as well as with 'Mtera dam', about river water use but do not consider coordination with the RBWO (Interviews 10 and 15).

Whose discourse counts?

Depending on their national-level financial, political and institutional backing, individual district departments and the RBWO have different degrees of upward or downward accountability, different leeway, and therefore different allegiances and feelings of responsibility. These translate themselves in different reasons to claim or disclaim responsibilities in responding to climate change.

The struggles over water resources have to be situated in a broader context of ongoing struggles about local level discretionary decision-making powers. Tanzania's national development vision hinges on agricultural-led growth with new flagship policies announced in rapid successive order in recent years: MKUKUTA II, Vision 2025, Kilimo Kwanza (Agriculture First) and lately Big Results Now (Tanzania Planning Commission, 1999; URT, 2009; Tanzania Ministry of Finance and Economic Affairs, 2010). Although districts in theory have advanced decision-making powers in all areas, these policies heavily influence rural districts' actions and district agricultural departments far outweigh other policy domains in terms of allocated human and financial resources. District agricultural activities are almost exclusively funded through centrally transferred sectoral block grants (Iringa District Council, 2013). Budget planning and allocation in core policy domains such as agriculture should in theory be informed by the outcomes of the O&OD process, but in practice rigid and top down, inflexible annual sectoral grant budget guidelines and detailed spending limits prioritize achieving poverty reduction, socio-economic development and food security through agricultural-production-led growth in line with national flagship policies (Iringa District Council, 2013). Even though these policies show some recognition of climate issues, at the moment they do not provide the districts with official budget lines or categories for climate-related spending and districts have very limited discretion to reallocate national funds (Interviews 15 and 16). In a context where newer and grander agricultural growth initiatives and strategies are repeatedly being declared, upwardly accountable agricultural bureaucrats hence have to deliver on political promises and neither have the leeway nor motivation for introducing (adaptation) activities that do not fit with national 'guidelines' and 'priorities'.

Contrary to the district agricultural departments, the locally-funded district departments and the RBWO have the motivation to claim adaptation. With considerably less political and financial backing at the national level, the Departments for Planning, Community Development and Natural Resources have limited influence over the district's development path. Officers in these departments are therefore constantly looking for opportunities to raise district domestic revenue, increase their working budget and ultimately break down the agricultural hegemony. Attracting international funding for adaptation activities provides an avenue through which these officers potentially could achieve this goal (Interview 12). Officers moreover claim to feel downwardly accountable. In fact, the

Planning, Community Development and Natural Resources Department seem to have developed a common long-term strategic perspective of environmental, economic and socially sustainable local development based on stimulating and facilitating a locally diversified economy and supporting household income diversification, claimed by several officers to be in line with an adaptation agenda (Interviews 6, 8, 12, 14, 15 and 16). An example of this sustainability alliance can be found in these three departments' joint initiative for the establishment of a tourism committee at Iringa Council (Interview 14). Coordinated by an officer from the Community Development Department, this committee aims to advance and support the development of and investments in a local eco-tourism industry with the triple goal of environmental conservation, creating alternative household income sources and generating domestic revenue (Iringa District Council (2010); Interviews 14, 15 and 16).

Apart from considering adaptation responsibilities are part of its core mandate to ensure sustainable long-term water resources use, the RBWO sees claiming the adaptation agenda as a way to gain authority in relation to other government actors (Interview 23). Funded directly under the Water Ministry, the RBWO finds itself outside the hierarchical government structure and lacks higher level political backing when it comes to conflicts with district departments that depend directly on more powerful line ministries such as agriculture. Moreover, the RBWO's interest in this respect has to be situated not only against the background of struggles with district councils, but also with local and national politicians claiming water resources for their constituencies and for commercial agricultural investments (Interview 23).

At the time of fieldwork, no official adaptation activities had been planned or undertaken on the ground. However, some of the dynamics described in the previous sections point towards promising avenues for adaptation implementation at district level. Even though only a relatively small and in fact decreasing share of Tanzania's climate-change-relevant national spending is controlled at the regional and local government level (Yanda *et al.*, 2013), districts have some leeway with their domestically-generated revenue and our results show that departments depending on this domestic revenue are forming alliances that could be highly conducive to implementing adaptation activities on the ground. The historical underfunding of the natural resources, but especially the community development department, mean that these departments have a track record in making the most of limited resources. Indeed, in finding useful ways to support the agricultural department's activities in the field the community development department in Iringa District has managed to secure additional resources, such as a vehicle for its department (Interviews 14 and 15). Moreover, as exemplified by the establishment of the Tourism Committee at Iringa District council, this districts' Community Development Department acts as a true broker in coordinating an integrated approach to implementing new strategies at district level.

Discussion and Conclusion

The importance of enhancing institutional capacity to address the challenges of climate change adaptation at the district level and a multilevel governance framework as a way of advancing proactive adaptation and overcoming barriers are increasingly argued (Amundsen *et al.*, 2010). In sub-Saharan Africa, which faces a combination of challenges including climate change, this is even more urgent. The literature argues, however, that adaptation in this region often becomes more about positioning and power struggles than about actual adaptation practices. This chapter's recount of district bureaucrats' struggling with the emerging adaptation agenda in rural Tanzania is a clear case in point. In Kilolo and Iringa District councils, considering adaptation mandates is as much about gaining political influence and authority over other government actors as about identifying actual adaptation needs and integrated approaches to dealing with them.

In recent discussions on how to best analyse, conceptualize and frame observed successes, delays or failures in adaptation governance, the process of politization at different governance levels emerges as a key focus area (Biesbroek *et al.*, 2014; Eisenack *et al.*, 2014, 2015). A central point in this literature is an emphasis on the need for more actor-centred research with a focus on the actual processes of decision making in policy formulation and implementation (Eisenack *et al.*, 2015). Taking such an approach, our results clearly illustrate how local-level bureaucrats engage in and try to shape the processes of decision making but that these are often unrelated to knowledge about the actual climate risks at hand. Although this might be due to the fact that Iringa region, at least compared to other regions in Tanzania, has not yet experienced severe impacts of climate change, our results tend to support the general trend in the literature that adaptation is as much about the opening up of a new place to contest dominant discourses and funding regimes as it is to adapt to future climatic threats. One particularly encouraging finding is that adaptation discussions in Tanzania in some ways present local district officers with an opportunity to contest established power relations. Like many other African countries, Tanzania's national climate policy discourse has been accused of constructing an 'anti-politics of adaptation' where technocratic formulations of climate impacts and adaptation strategies alienate national level agendas from local realities (Smucker *et al.*, 2015). Smucker *et al.* (2015) argue that the purposeful non-addressing of locally important issues such as equity, fairness and justice in adaptation policies frees the way to recruit proposed 'technical' adaptation activities for the countries' overriding (large-scale) agricultural-development-led growth strategy. Our results illustrate that within Tanzania's district administrations several departments and agencies are starting to form alliances to counter this agricultural hegemony. Understanding adaptation to climate change in sub-Saharan Africa requires that we understand how actors engage with adaptation discourses and policies and for what reasons. Climate policies and implementations are not simply a matter of best-suited practices and policies and overcoming barriers to implementing them, but rather of who has managed to position themselves in the right place once, and of course if, money and policies begin to flow down into the system. In sub-Saharan Africa where local government is often a mix of upwardly and downwardly accountable departments with different resources, powers, alliances and feelings of responsibility (Bierschenk and Olivier de Sardan, 2014), this insight is even more relevant than in developed countries situated in different policy contexts. Our results therefore also show that if we want to implement successful adaptation strategies for local communities we need to carefully consider the relative authority and negotiating power, but also the brokerage potential of local bureaucrats in a precarious political environment in which climate change adaptation becomes one amongst other arenas through which job survival can be achieved.

The insights in this chapter have important implications for understanding potential challenges and opportunities regarding climate change adaptation and, in turn, for decisions regarding potential allocations of climate-change-adaptation funds particularly in sub-Saharan Africa. Understanding climate change adaptation as more than a technical challenge related to mapping vulnerabilities and projecting future climate change provides insights into how power struggles and manoeuvring potentially undermine, or indeed facilitate, adaptation.

Acknowledgement

The research was conducted in the framework of the DANIDA funded CLIVET project.

References

Adger, W.N., Arnell, N.W. and Tompkins, E.L. (2005) Adapting to climate change: perspectives across scales. *Global Environmental Change – Human and Policy Dimensions* 15, 75–76.

Agrawal, A., Perrin, N., Chhatre, A., Benson, C.S. and Kononen, M. (2013) Climate policy processes, local institutions, and adaptation actions: mechanisms of translation and influence (vol 3, p. 565, 2012). *Wiley Interdisciplinary Reviews – Climate Change* 4, 72–72.

Aldunce, P., Beilin, R., Howden, M. and Handmer, J. (2015) Resilience for disaster risk management in a changing climate: Practitioners' frames and practices. *Global Environmental Change – Human and Policy Dimensions* 30, 1–11.

Amundsen, H., Berglund, F. and Westskog, H. (2010) Overcoming barriers to climate change adaptation a question of multilevel governance? *Environment and Planning C-Government and Policy* 28, 276–289.

Arnall, A., Kothari, U. and Kelman, I. (2014) Introduction to politics of climate change: discourses of policy and practice in developing countries. *Geographical Journal* 180, 98–101.

Artur, L. and Hilhorst, D. (2012) Everyday realities of climate change adaptation in Mozambique. *Global Environmental Change – Human and Policy Dimensions* 22, 529–536.

Berrang-Ford, L., Ford, J.D. and Paterson, J. (2011) Are we adapting to climate change? *Global Environmental Change – Human and Policy Dimensions* 21, 25–33.

Berrang-Ford, L., Ford, J.D., Lesnikowski, A., Poutiainen, C., Barrera, M. and Heymann, S.J. (2014) What drives national adaptation? A global assessment. *Climatic Change* 124, 441–450.

Betsill, M.M. and Bulkeley, H. (2006) Cities and the multilevel governance of global climate change. *Global Governance* 12, 141–159.

Biesbroek, G.R., Termeer, C.J.A.M., Klostermann, J.E.M. and P. Kabat (2014) Rethinking barriers to adaptation: mechanism-based explanation of impasses in the governance of an innovative adaptation measure, *Global Environmental Change* 26.

Bierschenk, T. and Olivier de Sardan, J.-P. (eds) (2014) *States at Work. Dynamics of African Bureaucracies*. Brill, Leiden, The Netherlands.

Bulkeley, H. and Kern, K. (2006) Local government and the governing of climate change in Germany and the UK. *Urban Studies* 43, 2237–2259.

Cashmore, M. and Wejs, A. (2014) Constructing legitimacy for climate change planning: A study of local government in Denmark. *Global Environmental Change – Human and Policy Dimensions* 24, 203–212.

Conway, D. and Mustelin, J. (2014) Strategies for improving adaptation practice in developing countries. *Nature Climate Change* 4, 339–342.

Conway, D. and Schipper, E.L.F. (2011) Adaptation to climate change in Africa: Challenges and opportunities identified from Ethiopia. *Global Environmental Change – Human and Policy Dimensions* 21, 227–237.

Crabbe, P. and Robin, M. (2006) Institutional adaptation of water resource infrastructures to climate change in Eastern Ontario. *Climatic Change* 78, 103–133.

Division of Environment (2012) *National Climate Change Strategy*. Vice President's Office, Dar es Salaam, United Republic of Tanzania.

Eisenack, K., Moser, S.C., Hoffmann, E., Klein, R.J.T., Oberlack, C., Pechan, A., Rotter, M. and Termeer, C.J.A.M. (2014) Explaining and overcoming barriers to climate change adaptation. *Nature Climate Change Perspective* 4, 867–872.

Eisenack, K., Moser, S.C., Hoffmann, E., Klein, R.J.T., Oberlack, C., Pechan, A., Rotter, M. and Termeer, C.J.A.M. (2015) Reply to 'Opening up the black box of adaptation decision-making'. *Nature Climate Change* 5, 494–495.

Eriksen, S.H., Nightingale, A.J. and Eakin, H. (2015) Reframing adaptation: The political nature of climate change adaptation. *Global Environmental Change* 35, 523–533.

Granderson, A.A. (2014) Making sense of climate change risks and responses at the community level: A cultural-political lens. *Climate Risk Management* 3, 55–64.

Greene, S. (2015) Resilience Buidling in Tanzania: Learning from experiences of institutional strengthening. Review of learning and early contributions to Climate Resilient Development. *IIED Working Paper*. IIED, London.

Heinrich Böll Stiftung and ODI (2015) Climate Funds Update [Online]. Available at: www.climatefundsupdate.org (accessed 4 January 2017).

Hulme, M. (2008) Geographical work at the boundaries of climate change. *Transactions of the Institute of British Geographers* 33, 5–11.

IPCC (2014) Climate Change 2014: Impacts, Adaptation and Vulnerability. Part A: Global and Sectoral Aspects. In: Field, C., Barros, V., Dokken, D., Mach, K., Mastrandrea, M.D., Bilir, T., Chatterjee, M., Ebi, K.L., Estrada, Y.O., Genova, R.C. *et al.* (eds) *Contribution of Working Group II to the Fifth Assessment Report of the Intergovernmental Panel on Climate Change*. Cambridge, UK and New York, NY.

Iringa District Council (2010) *Medium Term Strategic Plan* 2011/12 to 2015/16. Iringa Rural District Council, Iringa, Tanzania.

Iringa District Council (2012) *Iringa District Profile*. Iringa Rural District Council, Iringa, Tanzania.

Iringa District Council (2013) *Iringa Rural District Council Socio-economic Profile 2013*. Iringa Rural District Council, Iringa, Tanzania.

Lesnikowski, A.C., Ford, J.D., Berrang-Ford, L., Barrera, M. and Heymann, J., (2015) How are we adapting to climate change? A global assessment. *Mitigation and Adaptation Strategies for Global Change* 20, 277–293.

McSweeney, C., New, M. and Lizcano, G. (2010) UNDP Climate change country profiles: Tanzania. Available at: http://www.geog.ox.ac.uk/research/climate/projects/undp-cp/UNDP_reports/Tanzania/Tanzania.hires.report.pdf (accessed 11 April 2017).

Ministry for Agriculture (2013) Tanzania Agriculture Climate Resilience Plan 2014–2019. Available at: http://www.fao.org/climatechange/42077-07d3c561d911f22c7a7d12d0bdf123dc0.pdf (accessed 4 January 2017).

Ministry of Water and Irrigation (2014) Rufiji Basin. Available at: http://www.maji.go.tz/?q=en/content/rufiji-basin (accessed 4 January 2017).

Naess, L.O., Newell, P., Newsham, A., Phillips, J., Quan, J. and Tanner, T. (2015) Climate policy meets national development contexts: Insights from Kenya and Mozambique. *Global Environmental Change* 35, 534–544.

Nalau, J., Preston, B.L. and Maloney, M.C. (2015) Is adaptation a local responsibility? *Environmental Science & Policy* 48, 89–98.

Nam, N.P., Trang, T.T.T., Tyler, S., Anh, N.Q., Sinh, B.T., Huy, N.N., Khanh, P. and Huong, D.T. (2015) Local planning for climate adaptation: Vietnam's experience. *Asian Cities Climate Resilience Working Paper Series*. IIED, London.

Pasquini, L., Cowling, R.M. and Ziervogel, G. (2013) Facing the heat: Barriers to mainstreaming climate change adaptation in local government in the Western Cape Province, South Africa. *Habitat International* 40, 225–232.

Paterson, M., Humphreys, D. and Pettiford, L. (2003) Conceptualizing global environmental governance: from interstate regimes to counter-hegemonic struggles. *Global Environmental Politics* 3, 1–10.

PMO-RALG (2009) *Local Government and Reform Programme II*. Tanzanian Prime Minister's Office for Regional and Local Government, Dodoma, Tanzania.

Ribot, J.C., Lund, J.F. and Treue, T. (2010) Democratic decentralization in sub-Saharan Africa: its contribution to forest management, livelihoods, and enfranchisement. *Environmental Conservation* 37, 35–44.

Sherman, M.H. and Ford, J. (2014) Stakeholder engagement in adaptation interventions: an evaluation of projects in developing nations. *Climate Policy* 14, 417–441.

Smucker, T.A., Wisner, B., Mascarenhas, A., Munishi, P., Wangui, E.E., Sinha, G., Weiner, D., Bwenge, C. and Lovell, E. (2015) Differentiated livelihoods, local institutions, and the adaptation imperative: Assessing climate change adaptation policy in Tanzania. *Geoforum* 59, 39–50.

Tanzania Ministry of Finance and Economic Affairs (2010) *MKUKUTA II. Tanzania: Poverty Reduction*. Strategy Paper NSGRP II, United Republic of Tanzania.

Tanzania Planning Commission (1999) The Tanzania Development Vision 2025. Available at: http://www.mof.go.tz/mofdocs/overarch/Vision2025.pdf (accessed 4 January 2017).

Taylor, M. (2015) *The Political Ecology of Climate Change Adaptation: Livelihoods, Agrarian Change and the Conflicts of Development.* Routledge, London and New York.

Tompkins, E.L. and Adger, W.N. (2005) Defining response capacity to enhance climate change policy. *Environmental Science & Policy* 8, 562–571.

URT (2009) *Kilimo Kwanza: Mapinduzi ya Kijani*. The Ministry of Agriculture, Food Security and Cooperatives, Dar es Salaam, Tanzania.

Urwin, K. and Jordan, A. (2008) Does public policy support or undermine climate change adaptation? Exploring policy interplay across different scales of governance. *Global Environmental Change – Human and Policy Dimensions* 18, 180–191.

Vogel, B. and Henstra, D. (2015) Studying local climate adaptation: A heuristic research framework for comparative policy analysis. *Global Environmental Change – Human and Policy Dimensions* 31, 110–120.

Wagner, G. and Zeckhauser, R.J. (2012) Climate policy: hard problem, soft thinking. *Climatic Change* 110, 507–521.

Weisser, F., Bollig, M., Doevenspeck, M. and Muller-Mahn, D. (2014) Translating the 'adaptation to climate change' paradigm: the politics of a travelling idea in Africa. *Geographical Journal* 180, 111–119.

Yanda, P.Z., Mushi, D., Henku, H.I., Maganga, F., Minde, H., Malik, N., Kateka, A., Bird, N. and Tilley, H. (2013) *Tanzania National Climate Change Finance Analysis.* Overseas Development Institute, London, and the Centre for Climate Change Studies, University of Dar es Salaam, Tanzania.

MINISTRY OF LOCAL GOV'T.
HOUSING, EARLY EDU. AND ENV. PROTECTION
KAZUNGULA DISTRICT COUNCIL.
P.O Box 61133, LIVINGSTONE.
Tel-Fax : 021-323881.
Tel : 021-324835.
CIVIC CENTRE

4 Climate Change Adaptation and Decentralisation Politics: the Case of Local Governments in Rural Zambia

Mikkel Funder, Carol Mweemba and Imasiku Nyambe

Introduction

The COP 21 global climate change agreement in Paris in 2015 set a new direction in global agreements by emphasizing a 'bottom-up' country-driven approach to global climate action, based on voluntary Intended Nationally Determined Contributions (INDCs). The agreement thereby positioned national governments and other domestic actors as key drivers of global climate action. Alongside and in extension of this, local governments and other subnational actors have increasingly come into focus as key players in addressing climate change. In adaptation terms, this is partly because local governments tend to be the level of government most directly confronted with the everyday impacts of climate change, as they are forced to grapple with damaged roads, flooding rivers or drought-stricken voters. More fundamentally, local governments operate in the political space between national governments and communities, thus playing an important, if often ambiguous role as facilitators of national policies and representatives of local citizens.

So far, much of the emphasis on the role of local governments in climate change has evolved around *urban* local governments; there has been much less attention to rural local governments, especially in the context of developing countries. The growing recognition of local governments in addressing climate change has therefore, to a large extent, been driven by the world's big cities that have been well represented at UN negotiations and in the media as the 'first movers' on climate change. This makes good sense in many ways: many of the world's emissions emanate from the world's large cities, and with their economic and political punch they are important drivers of local government influence and action on the global climate agenda through efforts such as the C40 (Cities Climate Leadership Group) initiative and the ICLEI (Local Governments for Sustainability) network.

In the context of adaptation in developing countries, however, rural local governments are also important players. Despite urbanization, a large proportion of Africa's population continues to live and sustain their livelihoods in rural settings, and they are being hard hit by climate change. With decentralization policies in place in many African countries, rural local governments are, in principle, well positioned to create an enabling environment for adaptation and supporting specific activities (Sharma *et al.*, 2014). But what is, in fact, the situation on the ground? What is the real-world context in which rural local governments in African countries operate in terms of climate adaptation, what are the conditions and constraints they face, and how are they responding?

This chapter contributes to our understanding of these issues through a discussion of selected findings from two rural districts in Zambia, drawing on research conducted in 2012–2016 as part of the research programme on Climate Change and Rural Institutions (Christoplos *et al.*, 2016). Our focus is less on specific capacity and funding issues, and more on the underlying institutional and political dynamics of which rural local governments form part. In particular we discuss how the relationship between the central state and local governments and the associated politics of decentralization influence subnational climate change adaptation.

Zambia's National Adaptation Goals

Multilateral and bilateral donors have played a major role in the emergence of Zambia's climate change agenda and associated institutional frameworks. This does not mean, however, that climate change is merely an exogenous interest. Indeed, the climate change agenda supports a variety of broader political interests among Zambia's ruling elites. In terms of adaptation, this includes securing funding for infrastructural development and rural extension services, as well as ensuring voter support and avoiding public protests related to food insecurity and food prices (Funder *et al.*, 2013).

These interests are also reflected in Zambia's intended commitments under the Paris Agreement. The country's INDC emphasizes 'guaranteed food security through diversification and promotion of Climate Smart Agricultural (CSA) practices' as a priority in adaptation (GRZ, 2015, p.7). Other priority areas include institutional development and capacity-building in water management, land-use planning and infrastructure development, and improved decentralized climate information services. These priorities reflect similar objectives in Zambia's general national development plans and build on other climate policy and strategy documents, including the National Adaptation Programme of Action (NAPA) of 2007 (GRZ, 2007), the National Climate Change Response Strategy of 2010 (GRZ, 2010), and the draft National Policy on Climate Change. The INDC estimates that US$20 billion are required for adaptation actions up to 2030, and stresses that Zambia's commitments to both adaptation and mitigation are contingent on international financing and support.

In institutional terms, Zambia's national climate strategies and plans on climate change adaptation have so far focused mainly on the development of national frameworks and ministerial capacities. The country's long-standing decentralization policy has only been perfunctorily addressed in climate plans and the role of local governments in climate adaptation has been unclear. Recently, however, there have been some signs of greater attention being given to the role of decentralized adaptation mechanisms among national climate policy makers. A major influence in this respect has been the World Bank-coordinated Pilot Program for Climate Resilience (PPCR), which includes a portfolio of activities to enhance decentralized planning for climate change in policy and practice (see below). Zambia's INDC thus goes some way towards acknowledging decentralized adaptation:

> Planning for adaptation and mitigation programs under Zambia's INDC will be integrated in existing planning processes and supported by national budget allocations to sectors, ministries and sub-national authorities [...] In addition, the decentralisation process currently being undertaken in the country will enhance multi stakeholder participation in the implementation of the INDC.
>
> (GRZ, 2015, p.11)

But what are the de facto options, constraints and political dynamics of such decentralized climate governance in practice? In the following section, we explore this by examining the real-world context of decentralization and subnational adaptation in Zambia.

The Politics of Decentralization

Upon independence, Zambia inherited the colonial administrative system whereby so-called Native Authorities, spearheaded by chiefs appointed by the colonial authorities, formed the main institutionalization of 'indirect rule' (Hobsbawm and Ranger, 1983). The Native Authorities were under the control of district and provincial authorities, who were supported by technical government officers in the relevant sectors. Some aspects of this system continue today, albeit in a different form. Chikulo (2009) speaks of four major phases in Zambia's decentralization following independence (see also Tordoff and Young, 1994; Mukwena, 2001; Brown, 2005; Battera, 2016):

- During the first phase from 1964 to 1970, the newly independent government of President Kenneth Kaunda sought to assert its control over rural areas. This included abolishing the Native Authorities – a key platform for chiefly authority – and introducing instead a new local government structure. In practice, this was controlled from the centre through district governors acting on behalf of the president, as well as deconcentrated technical line agencies who continued operating much as under colonial rule.
- A second phase from 1971 to 1979 focused on developing a subdistrict grassroots structure in the form of elected ward and village committees as a means to provide bottom-up planning and representation in district-level planning and development. Alongside this, Kaunda's United National Independence Party (UNIP) built an effective party structure in rural areas around which local politics evolved in practice.
- In the third phase from 1980 to 1990, any pretence at distinguishing between local government and the UNIP party structure was abandoned in favour of a fully integrated system whereby the party had exclusive control of the local government mechanism. Non-party members could not stand for election to local government, and district councils were formed in which local government representatives, the party and central government agencies coordinated development efforts.
- The fourth phase from 1991 into the 2000s reflected the end of UNIP rule and the introduction of multi-party democracy. The party and district councils were delinked, and free local government elections introduced. A by-product of these changes was that the village and ward development committees effectively became dysfunctional in most districts, leading to a gap in the subdistrict governance structure, which remains to this day.

As an extension of the latter developments, a decentralization policy was drawn up in 2002, which in principle aimed at devolving functions and authority to district councils in a number of areas, including development planning, agricultural extension, disaster management, environmental services, water and land allocation. The policy was nevertheless careful to emphasize that:

> When the management capacity has been built, Councils will be given more authority and higher degree of autonomy while maintaining sufficient linkages with the centre as demanded in unitary state. However, this is a long term task requiring an evolutionary approach.
>
> (GRZ, 2002, p.22)

This somewhat restrained approach was reflected in implementation over the ensuing decade, during which time little actual progress in decentralization was made. In 2013, the existing policy was updated and slightly revised under the populist rule of the late President Sata, including a greater emphasis on grassroots inclusion through reinstatement of subdistrict ward development committees. Since then, some measures towards actual implementation have been carried out. This includes preparing and endorsing the legal and administrative basis for decentralization and an overall action plan of step-wise devolution of particular functions during the period 2015–2017, including agricultural extension and disaster management. So far, activities have mainly evolved around capacity development, typically in the form of training by central government staff and donor-supported expert assistance. Indeed, the updated policy of 2013 retains the earlier emphasis on a gradual devolution process in which it is up to the central government to determine when councils are ready to take on functions on their own. The planning and budgeting structure also remains subject to a laborious and bureaucratic approval process from higher levels.

Three key features thus characterize Zambia's decentralization process to date:

- Firstly, it has been highly politicized and tied to the changing national political discourses, most obviously in the linking of local government and party structure under UNIP, but also evident in the neoliberal dismantling and partial disruption of subdistrict structures in the post-UNIP years, as well as the populist emphasis on devolution more recently.

- Secondly, throughout these changes central government line ministries and agencies have remained the primary authorities and leaders in decision making, planning, implementation and extension within their respective fields. District-level agencies have thus been deconcentrated rather than devolved and despite recent momentum in the decentralization process this remains the case today, as will be discussed below.
- Thirdly, local government planning processes have been disrupted and restructured on several occasions and remain subject to centralized approval and control in many respects. Historically, the central level has thus been reluctant to pass over control of planning and budgets to the lower levels. In this respect, a 'lack of capacity' has been an oft-repeated discourse, initially as an argument for not devolving real power and funds, and latterly as a device for central government to control when and under what conditions district councils will be 'ready' to take on particular functions.

Importantly, the efforts of the central state to retain control and carefully steer the decentralization process should not be seen as an indication of a central state with monolithic power: following the fall of UNIP and the dismantling of the rural party structure, a variety of competing forces have (re-)emerged in rural areas, including chieftaincy structures, competing political parties and factions, business interests and civil-society organizations (Gould, 2010). The reluctance of the central state to carry out full devolution can be seen as an attempt to retain control in a context of political competition and limited state reach on the ground.

In the following, we discuss the implications of this context for local governments in relation to climate adaptation and how they are responding to it through the case of two districts in south-western Zambia.

Institutional Responses to Climate Change in Kazungula and Sesheke

The neighbouring districts of Kazungula and Sesheke are located in Zambia's Southern and Western Provinces, respectively, in the country's driest region along the Zambezi river. The districts are highly rural, and infrastructure is limited. The main ethnic groups are the Lozi, Tonga and Toka Leya, whose livelihoods revolve around rain-fed agriculture, with maize as the dominant crop, as well as livestock – especially cattle – and artisanal riverine fisheries. Poverty rates are high, and in some areas more than 60% of the population are categorized as food-insecure. The districts are located in what is considered the most climate-impacted agroecological region of Zambia (GRZ, 2007). Recent decades have seen average temperature increases, as well as changing rainfall patterns towards shorter rainy seasons. This has especially affected maize production because current varieties are unable to cope. At the same time, a series of extreme floods of the Zambezi and its tributaries throughout the 2000s have been attributed to climate change and have severely disrupted local livelihoods. The impacts of climate change are also interacting with a range of other factors in the area, including livestock disease and increasing competition over fishery resources and land.

Local institutional actors and interests

The local institutional landscape in the two districts is multi-faceted, and a range of actors is involved formally or informally in climate change adaptation efforts. In the following, we briefly discuss these actors and their adaptation activities, with an emphasis on central and local government.

Central government is represented in the two districts through the deconcentrated district-level structure. This includes the district administration, which is headed by the district commissioner (DC), who is appointed by central government and is therefore de facto a political position. The district line agencies are deconcentrated units of their respective parent ministries. They are typically under-capacitated compared to the policies and mandates they are supposed to implement and enforce, a situation worsened by the structural adjustment requirements of the International Monetary Fund (IMF), which resulted in heavy cuts to the civil service (Duncan *et al.*, 2003). Deconcentrated state agencies therefore have a strong interest in asserting and improving their role and presence through extension activities in adaptation and generally.

Local government in Kazungula and Sesheke takes the form of district councils, consisting of the council itself and its executive sections. Councils consist of elected councillors, members of parliament (MPs) and two representatives appointed by the chiefs in each district. Councillors are democratically elected for 5-year periods through ward constituencies. District councils form the local representative bodies in the district planning process and are also charged with the responsibility for delivering selected public services. These include some aspects of water and sanitation, land-use planning, feeder road development and the development of urban infrastructure and housing. District councils are funded partly through the annual government budget of the Ministry of Local Government and Housing and partly through revenues generated from local resources. In practice, the latter are typically very limited because key natural resources in their area tend to be controlled by the central state (e.g. timber and minerals) or by the chiefly structures, which control communal land. The district councils therefore have a very limited revenue base and are critically dependent on central government budget allocations. In terms of technical capacity, they are, to a large extent, dependent on the technical line agencies of central government but have no direct authority over them, so are as such, to a large extent, dependent on their goodwill in everyday practice.

A variety of other institutional actors are involved formally and informally in adaptation activities in Kazungula and Sesheke. This includes chieftaincy structures, members of parliament, non-governmental organizations (NGOs) and grassroots organizations. In a context of legal pluralism and institutional multiplicity, these actors compete for leverage and authority, and some – such as the chiefs – often directly contest the legitimacy of the central state. We have discussed this in more detail elsewhere (Funder *et al.*, 2017). Significant to note here is the particularly active role of NGOs in implementing adaptation projects.

Who carries out adaptation?

In formal terms, the main fora for planning and coordinating adaptation and disaster management activities at the district level are the various District Coordination Committees, most notably the District Development Coordinating Committee and the District Disaster Management Committee. These include representatives from the district administration and the state line agencies, whereas the district council is represented by the executive section of the council and selected councillors. Representatives of chiefs and NGOs are usually also present.

In practical terms, however, adaptation activities have so far been mainly carried out as project-specific activities with their own timelines and budgets, or as ad hoc emergency responses to floods and droughts. The large majority of activities has therefore been both implemented and dominated by central government agencies and NGOs.

Central government activities have primarily been led by two agencies. Operating under the Vice President's Office, the Disaster Mitigation and Management Unit (DMMU) has been a pivotal actor in addressing flood and drought responses in the area. This includes leading and coordinating immediate disaster responses, but in recent years the DMMU has also increasingly taken on a central role in longer-term adaptation responses in the district, including a model flood resettlement scheme dubbed a 'climate change adaptation' scheme. The other major government actor in adaptation has been the Ministry of Agriculture, which undertakes extension activities on adaptive practices such as conservation agriculture, early maturing maize varieties, drought-tolerant crops such as millet and sorghum, the introduction of rice in flooded areas and pilot micro-irrigation schemes. Other central government agencies, such as the Department of Water Affairs and the Meteorological Services, also have a more limited number of adaptation activities. Most adaptation activities by government agencies are financed through programmes and projects funded by multilateral or bilateral donors such as the United Nations Development Programme (UNDP), the Food and Agriculture Organization (FAO) and the World Bank, typically anchored in the parent ministries in Lusaka.

In addition to this, a number of climate-related projects have been carried out by international and national NGOs such as CARITAS, Red Cross, Save The Children, World Wide Fund for Nature (WWF) and the African Wildlife Foundation. These are almost universally conducted as pilot schemes for community-based adaptation and disaster management, with activities such as conservation agriculture, the introduction of more drought-resilient crops and livestock, grain banks, and the development of community-based fora and mechanisms. Funding for these activities has typically been derived from international sources.

By contrast, district councils have carried out few actual adaptation projects. Their involvement in climate and disaster management activities has thus tended to revolve mainly around their participation in the District Development Coordination Committees and the Disaster Management Committees and associated activities. Their formal role in the actual planning, steering and implementation of adaptation and disaster management activities has thus been limited. One recent exception to this has been the Pilot Programme for Climate Change Resilience (PPCR). Funded under the World Bank-managed Climate Investment Funds, this is the first major donor-funded effort in Zambia to seek to strengthen district and subdistrict adaptation planning. Anchored in the Ministry of Finance, the programme initially worked to advise national policy makers on the significance of climate adaptation, and in 2015 it began funding subnational adaptation activities with the involvement of local governments in a number of districts, including Kazungula and Sesheke. At the time of writing the programme has focused on subdistrict community-level adaptation activities, as well as the initial training of district council staff in, for example, the climate-screening of district development plans and making a limited amount of funding available for council-led adaptation projects.

With the exception of the PPCR programme, local governments have thus, to a large extent, been sidelined in district-level adaptation activities so far. This does not arise from a lack of interest, but reflects the structural position of local governments and, in particular, their relationship to central government. This is discussed in the following section.

The Persistence of Centralized Control in Adaptation

Apart from the formal district-level coordination fora, technical staff from the state line agencies interact on an everyday basis with planners and administrative staff in the executive arm of the district council. In this sense, the central state is not 'remote' from district councils. Personal relationships are commonplace, as are pragmatic approaches to solving crises. Although local staff in central government agencies typically have amicable relations with councils, they are, however, also under direction from their mother ministries at the central level, and eager to protect and boost their own technical and territorial authority. The everyday interactions between central government staff and district councils in adaptation therefore play out within the broader picture of central state efforts to retain control. This has several implications.

Firstly, district councils perceive a major disconnect between national-sector policies and plans and the de facto needs of councils in responding to climate adaptation. National climate-related sector plans are seen by councils as one-size-fits-all documents that dictate rather than facilitate local responses to adaptation. In Sesheke District, for example, the state district agricultural office provides extension services for conservation agriculture as prescribed in the national agricultural policy and the National Adaptation Action Plan. Farmers in the area complain, however, that the sandy soils of Sesheke are not conducive to the methods advised and that extension should focus on alternative adaptation techniques. Councillors and administrative staff have discussed the matter with government agricultural officers, who acknowledge the problem but are tied down by ministerial plans and requirements. Also in Sesheke, government agricultural officers have proposed a collaborative project on water-harvesting and small-scale irrigation projects that would include district councils as collaborating partners. However, although the district agricultural officer has repeatedly proposed these activities in Annual Workplans and Budgets, they have consistently been turned down at the provincial and central levels as being beyond policy priorities.

Secondly, central government has been reluctant to devolve and relinquish de facto control over decision making and financing to lower levels. This includes core budgets for salaries and operational activities in district councils, which are often delayed or below approved budgets.

Climate-specific funding in donor-supported programmes is also affected. For example, a multilaterally financed programme aimed at facilitating extension on adaptive agricultural practices in several districts in southern Zambia, including Kazungula, suffered delays of up to 2 years because the central-level Ministry of Agriculture required the planned activities to be carefully scrutinized and aligned with national policies before releasing funds. According to provincial and district agricultural staff, one implication of this was that the programme concentrated on meeting national objectives and refrained from engaging with broader district-level planning.

Reluctance to approve plans and disburse funding is thus not only a function of cumbersome bureaucracy, but reflects an emphasis at the central level on retaining control of the nature and financing of adaptation measures carried out on the ground. As mentioned above, following the revised decentralization policy of 2013 there has been some progress in moving towards a greater emphasis on actual devolution, yet autonomous decision making and financial control of adaptation measures remains limited. One example of how centralized control persists under nominal devolution is evident in the PPCR programme, which, despite its relatively progressive approach, has been locked into conventional planning and decision-making structures: within the programme, approval of even a small-scale community adaptation project in Kazungula or Sesheke thus needs to pass first through an NGO contracted by central government, after which it requires approval from the District Development Coordination Committee, followed by approval from a provincial coordination unit, and eventually from a national ministerial unit, which may choose to send out its own mission to verify the project. Hence, although rural districts are here involved in decision making, their influence is watered down by a number of other subsequent decisions at higher levels.

Thirdly and finally, rather than surrender their authority in agriculture, water and other adaptation-related areas, central state agencies in Sesheke and Kazungula have so far worked actively to consolidate and extend it. One example of this has been the state's responses to a series of major floods in the two districts during the 2000s. Although originally only charged with coordinating and supporting immediate emergency relief efforts, the DMMU has increasingly expanded its mandate to become a dominant force in climate change adaptation in Kazungula and Sesheke more broadly. This has included driving the development of a flood resettlement scheme and associated land and agricultural development activities in Kazungula, with the longer-term aim of opening up territory to government control and making land and production systems subject to technical interventions by state agencies (Funder *et al.*, 2017). While initially positive towards these activities, the Kazungula District Council eventually became wary of their top-down nature and the local conflicts they have generated.

Local Government Responses

How are local governments responding in this context of the centralized control of adaptation activities? In overall terms, district councils manoeuvre pragmatically between, on the one hand, associating themselves with the state, and on the other hand distancing themselves from it and claiming autonomy. This two-pronged approach reflects a key interest of district councils, namely to assert their legitimacy in a context where – given the history of local governments in Zambia – they have come to be seen as weak and underfunded entities by both their constituencies and the state agencies with which they work. The fact that it is at all possible for district councils to manoeuvre when considering the force of the central state reflects the fragmented nature of the local institutional landscape, which means that the central state cannot completely disregard councils and that alternative alliances beyond the central state are a possibility. Specifically, the Kazungula and Sesheke District Councils have three strategies in this respect, as described next.

Engaging and 'piggy-backing'

Although district councils disapprove of their limited mandate and financing in adaptation, they do not sit back passively. On the contrary, elected councillors and executive staff engage actively in adaptation efforts whenever opportunities allow, including those led by central government agencies. In some cases this is an expected requirement that cannot be evaded, whereas in others it is a voluntary action. An example of the former is the emergency responses to flood disasters, where councils essentially act under the direction of the DMMU and the District Commissioner. Although councils are eager to gain more de facto influence in disaster management, they are well aware that crisis situations are a bad time to resist central government control. Participating in the emergency relief activities also affirms a role for councils as legitimate actors in disaster management in the first place. For example, in the aftermath of particularly extreme floods in the 2005–2006 rainy season, the Kazungula District Council played a key role in establishing an emergency camp near the council headquarters. Although this was done at the instruction of the DMMU, it helped position the council as a capable actor in disaster situations and was well regarded in the local constituency.

In other situations, councillors or executive staff engage in government adaptation activities even where they have no formal role. In Sesheke, councillors have actively become involved in micro-irrigation and water development projects led by the District Agricultural Office and Department of Water Affairs, offering their assistance. Given their role as elected local government representatives, the state agencies have had little choice but to accept this and involve them in planning meetings, during which councillors have then sought to influence the implementation and expansion of the projects. Such strategies also provide opportunities for personal political leverage: in both Sesheke and Kazungula, some Councillors have presented NGO-led projects as activities initiated on their behalf, despite having played no role in the origin of the projects.

Councillors and executive staff thus seek to gain influence on government adaptation projects by actively engaging in them and positioning the council and its members as legitimate players in disaster management and adaptation by 'piggy-backing' on the authority of the central state.

Seeking alternative alliances

On other occasions, district councils have sought to sidestep the domination of central government agencies in adaptation by seeking alternative alliances and sources of funding. For example, during the planning process for the above-mentioned PPCR project, it was originally intended to provide funding for a flood-resettlement scheme in Kazungula led by the DMMU. The district council, however, felt increasingly sidelined by the DMMU in this scheme and argued to a World Bank planning mission that the funding was better spent elsewhere in the district. As part of this argument, the council pointed to its position as knowing best about local adaptation needs. Despite the strong central-level control of the PPCR programme, the council's effort to convince the programme's coordinating donor agency proved successful and the funding was diverted to other areas.

In addition, recent efforts have been made by the two district councils to harness the resources and know-how of NGOs. Typically, NGOs working on adaptation in the area have paid little attention to district councils and have either implemented their projects independently or have sought to align them with central government line agencies. Seeking to address this situation, councils have sought to actively enforce a requirement that NGOs consult and plan all activities in close collaboration with themselves. Although some NGOs reject this as an attempt to control their work, others have begun to engage more with councils, especially in areas such as water and disaster management, which are in principle due for devolution under the 2013 decentralization policy.

Challenging central government

A final and less pragmatic approach has been evident in recent years, in which councillors have employed climate change as a platform for challenging the authority, efficiency and resource control of the central state. There are various manifestations of this. Firstly, some councillors have openly criticized state agencies for failing to address the impacts of climate change adequately. This includes certain defunct boreholes, which local communities and their councillors claim have been dried out by climate change. The Department of Water Affairs, for their part, argue that the problem is a result of poor community management of boreholes. Councillors have countered by arguing that the department is evading its responsibilities and failing to take climate impacts on local populations seriously, thereby implicitly criticizing the ability of central government to serve the needs of the people.

Secondly, and as an extension of this, the councils in the two districts employ climate change in discourses of devolution. During meetings with central government agencies and donors they argue that, because of their local and democratic nature, they are the best placed organization to address climate change. In one instance, the Chair of the Sesheke District Council wrote a letter directly to parliament in which he described recent floods and droughts, and argued that, because of climate change, devolution of funds and powers to district councils was now more important than ever.

Lastly, some councillors employ climate change as a platform for challenging the central state's control over land and other natural resources in the districts. In Kazungula, councillors have thus publicly argued that the area's State Forest Reserves and other conservation areas should be degazetted in order to provide land and alternative incomes from charcoal production for flood- and drought-affected communities. Climate change thus enters into existing conflicts over land and resource control in the area, giving councillors an opportunity to challenge established mechanisms of resource control.

Conclusion and Policy Implications

What do the cases of Kazungula and Sesheke Districts imply for subnational climate action more broadly? They show, first, that moving from the Paris agreement and national policies to action on the ground involves more than capacity and financing. These are important issues, but they cannot be addressed or understood outside the context of governance. Our study therefore illustrates how the ability of local governments to act on climate change evolves around the de facto politics of subnational governance, including the relationship between the central state and local institutions. National climate policies and efforts to support them must therefore engage more directly with the realities and challenges of decentralization. It is not enough merely to pay lip-service to existing decentralization policies or to express an intention to work through them, as in the Zambian case. Their nature and character must be understood and placed at the centre of climate adaptation policies. A first step in this direction can be to examine how subnational governance and decentralization processes influence climate action in the given country or setting, determining what needs to be done on this basis.

In particular, our study shows how national governments and their sector agencies may seek to monopolize climate action, rather than relinquish control to local governments and other local actors, despite decentralization policies being in place. A deterministic reading should, of course, be avoided. In Zambia, the recent steps to move forward with the general decentralization policy and the climate-specific PPCR programme show some promise, yet centralized control persists in key decision-making processes related to adaptation. Indeed, as our research in Zambia suggests, adaptation interventions are in some cases employed by the central government to *increase* centralized control over people and resources. The reluctance of national governments to follow through on devolution is well documented in decentralization studies more broadly (Boone, 2003; Crook, 2003; Ribot, 2007) but the implications for subnational climate action need more attention. This is particularly important given the historical tendency to place nation states at the heart of global climate agreements and finance mechanisms. While this is necessary and sensible in many ways, it carries an inherent risk that national governments abuse their role as national anchoring points and 'gatekeepers' in climate implementation and financing. Climate agreements and financing mechanisms such as the Green Climate Fund can play a role in this respect by incorporating measures that enhance the downward accountability of national governments and support platforms for local governments, civil society and other actors to counter autocratic adaptation (Funder, 2016).

Another key lesson from Kazungula and Sesheke is that local governments do not merely sit back and watch, but engage actively and strategically in adaptation efforts. This provides an important basis for climate adaptation support to build on and suggests a need to move beyond the common perception of Africa's rural local governments as inherently passive and incapable. Indeed, as the case of Zambia demonstrates, such discourses may serve to justify continued centralization in climate change adaptation. Clearly, capacity issues exist and must be addressed but the key to doing so is a matter of governance, that is, of ensuring that local governments are in a position to direct the technical inputs of local-sector agencies according to locally developed adaptation needs and plans. In this respect it is interesting to note that some staff in the deconcentrated government agencies in Kazungula and Sesheke are themselves becoming wary of centralized, one-size-fits-all policies emanating from their parent ministries. Frustrations over the disconnect between central-level policies and local needs is thus potentially a common future meeting ground between local governments and deconcentrated technical staff.

Our study further shows that naïve assumptions of 'good' and 'bad' actors in climate governance are misplaced. As shown above, personal leverage plays an important role when councillors and other actors engage in adaptation (cf. rent-seeking among district councillors). Accountability mechanisms are therefore as important for local governments as they are for central governments in adaptation (Mweemba, Funder and N'goma, unpublished). However, as this case also shows, climate change adaptation can provide a platform for local governments to assert their roles and claims regarding central governments more broadly. This includes forming new alliances between local governments and civil-society organizations, and staking claims for fiscal devolution and locally anchored natural resource governance. Strengthening subnational climate governance thus potentially has wider-reaching benefits for governance and resource control.

Addressing the challenges of subnational climate adaptation and the role of local governments therefore requires greater attention to the politics of decentralization, but it is not an impossible task. Emerging experiences from Kenya (Barrett, 2015) and Ghana (Chapter 6, this volume) provide initial guidance for the way forward and must be approached as more than isolated projects. Subnational climate governance is not an appendix to the global climate change agreement but a fundamental precondition for its success.

References

Barrett, S. (2015) Subnational adaptation finance allocation: comparing decentralized and devolved political institutions in Kenya. *Global Environmental Politics* 15, doi:10.1162/GLEP_a_00314.

Battera, F. (2016) State, poverty and agriculture in Zambia: the impact of state policies after democratization. In: Pallotti, A. and Tornimbeni, C. (eds) *State, Land and Democracy in Southern Africa*. Routledge, UK.

Boone, C. (2003) Decentralization as political strategy in West Africa. *Comparative Political Studies* 36, 355–380.

Brown, T. (2005) Contestation, confusion and corruption: market-based land reform in Zambia. In: Evers, S., Spierenburg, M. and Wels, H. (eds) *Competing Jurisdictions: Settling Land Claims in Africa*. Brill, Leiden, The Netherlands.

Chikulo, B. (2009) Local governance reforms in Zambia: a review. *Commonwealth Journal of Local Governance* 2, 98–106.

Christoplos, I., Aben, C., Bashaasha, B., Dhungana, H., Friis-Hansen, E., Funder, M., Huong, N., Khatri, D., Lindegaard, L., Mweemba, C. *et al.* (2016) Understanding sub-national climate governance: Findings from Nepal, Uganda, Vietnam and Zambia. *DIIS Report*. Danish Institute for International Studies, Copenhagen.

Crook, R.C. (2003) Decentralisation and poverty reduction in Africa: the politics of local–central relations. *Public Administration and Development* 23, 77–88.

Duncan, A., Macmillan, H. and Simutanyi, N. (2003) *Zambia: Drivers of Pro-poor Change: an Overview*. Report prepared for the Department for International Development by Oxford Policy Management, Oxford, UK.

Funder, M. (2016) Policy-makers and climate financing in African LDCs: the case of Zambia. In: Fejerskov, A., Funder, M., Engberg-Pedersen, L., Jiang, Y., Ravnborg, H. and Webster, N. (eds) *Financing Sustainable Development: Actors, Interests, Politics* (DIIS Report 2016:1). Danish Institute for International Studies, Copenhagen.

Funder, M., Mweemba, C.E. and Nyambe, I. (2013) The climate change agenda in Zambia: national interests and the role of development cooperation. *DIIS Working Paper 2013:13*. Danish Institute for International Studies, Copenhagen.

Funder, M., Mweemba, C., and Nyambe, I. (2017) The Politics of Climate Change Adaptation in Development: Authority, Resource Control and State Intervention in Rural Zambia. *Journal of Development Studies*. Available at: http://www.tandfonline.com/doi/full/10.1080/00220388.2016.1277021.

Gould, J. (2010) *Left Behind: Rural Zambia in the Third Republic*. Lembani Trust, Lusaka.

GRZ (2002) *The National Decentralisation Policy: Towards Empowering the People.* Government of the Republic of Zambia.

GRZ (2007) *Zambia National Adaptation Programme of Action.* Government of the Republic of Zambia.

GRZ (2010) *National Climate Change Response Strategy.* Government of the Republic of Zambia.

GRZ (2015) *Zambia's Intended Nationally Determined Contribution to the 2015 Agreement on Climate Change.* Government of the Republic of Zambia.

Hobsbawm, E.J. and Ranger, T.O. (1983) *The Invention of Tradition*. Cambridge University Press, Cambridge.

Mukwena, R.M. (2001) Situating decentralisation in Zambia in a political context. *African Administrative Studies* 57, 35–50.

Ribot, J. (2007) Representation, citizenship and the public domain in democratic decentralization. *Development* 50, 43–49.

Sharma, V., Orindi, V., Hesse, C., Pattison, J. and Anderson, S. (2014) Supporting local climate adaptation planning and implementation through local governance and decentralised finance provision. *Development in Practice* 24, 579–590.

Tordoff, W. and Young, R.A. (1994) Decentralisation and public sector reform in Zambia. *Journal of Southern African Studies* 20, 285–299.

Brevetto Internazionale

5 A White Elephant in a Changing Climate: a Territorial Approach to Climate Change Adaptation in Uganda

J.J. Okiror, Esbern Friis Hansen, Bernard Bashaasha and Isaac Nakendo

Introduction

This chapter examines the institutional sustainability of internationally financed support for rural climate change adaptation. The most common approach for such support is the *planned climate change adaptation approach*, which relies on projects as a mechanism for implementation and governance. This raises two concerns. Firstly, a project mode of implementation frequently establishes structures, procedures and funding mechanisms parallel to those of government, have limited time framework and have been widely criticized and abandoned in development work in favour of participatory planning and decentralized governance structures (Huppi, 2006).

Chapter 2 identified a lack of coherence between different administrative levels of government when it comes to how decisions over climate change funding priorities are made and where actions are needed. Central ministry staff and international donor agencies set the agenda for climate change adaptation priorities and control access to financial support, while the role of local government administration is limited to that of implementing agencies that contradict decentralization policies in many countries. This chapter examines the implications of this lack of fit between central and local government regarding how climate change adaptation actions are planned and implemented, as well as for the institutional sustainability of the actions taken.

The empirical case study chosen here is the joint local government–donor project in Uganda that applies the United Nations Development Programme (UNDP)'s Territorial Approach to Climate Change (TACC). The TACC project, implemented in 2012–2013, represents the most comprehensive attempt to date by international development partners in Uganda to address rural climate change adaptation at the local government level. The Ugandan TACC project provides a good example of planned climate change adaptation (PCCA) because it has received full support from the central ministries, national donor agencies and the UNDP. The TACC in Uganda is seen as a pilot project that aims to provide experience guiding future international investments in rural climate change adaptation.

In its original form, PCCA is a blueprint planning approach that takes its point of departure in science and expert knowledge (climatology) when it comes to identifying problems (Adger *et al.*, 2007). The TACC design seeks to address the democratic and participatory deficit that is inherent in the planned climate change approach by allowing popular consultation, engaging in instrumental participation and engineering new local institutions intended to represent the interests of the community (UNDP, 2013).

Overall this chapter seeks to explore the implications of the skewed funding modality for climate change activities (discussed in Chapter 2) for implementation at the subnational administrative levels. This project mode of implementation centralizes decision making at the ministerial and UN agencies level. However, adaptation measures can best be implemented at the district level in the context of decentralized local governments.

Study focus

This study does not seek to evaluate the TACC project in any way but instead uses TACC as an example with which to examine how national government and international development partners engage with existing decentralized governance structures in supporting climate change adaptation at the subnational level. Fieldwork took place in 2015, 2 years after the first phase of the TACC project had ended. More specifically, the study explores:

- The extent to which the parallel structures introduced by the TACC project have been integrated with existing local government institutions and procedures;
- The political and technical inclusiveness of the planning and implementation of the TACC project;
- The perception and sense of ownership of the TACC project among local government staff, political leaders and the rural population; and
- The extent to which activities initiated by TACC have been able to sustain themselves since the project ended.

A case-study approach was used to interact with local government leaders at district and sub-county levels. Key Informant Interviews and Focus Group discussions were held with stakeholders who were knowledgeable of the TACC activities in the three districts of Bududa, Manafwa and Mbale. They included: the Chief Administrative Officers, District Environment Officers, District Councillors and TACC beneficiaries. The focus group discussions and Key Informant Interviews were supported by field visits with observations of climate change effects and adaptation activities in the region.

Subnational Government and Implementation of Support for Climate Change Adaptation

In the early stages of development cooperation, development defined as aid to governments was provided through a project-funded intervention approach, influenced by specific implementation units (Cabral and Shepherd, 2008). Project-funded implementation was undertaken by donor agencies or organizations subcontracted to carry out specific project, these thus being the main actors responsible for design and implementation, as well as monitoring and evaluation of the projects (Cabral and Shepherd, 2008).

The transition from project-funded interventions to sector-wide approaches (SWAp) and basket funding was a result of the increased fragmentation experienced in development interventions owing to the proliferation of projects and funding agencies entering the development scene. SWAp and basket funding dealt with the inconsistencies that were apparent during the project-funded period, when different actors were not collaborating (Cabral and Shepherd, 2008). The Paris Declaration on Aid Effectiveness of 2005 consolidated these new approaches by adopting five main principles for donors and recipients, among them a requirement to produce harmonization through common funding and granting national ownership of projects (ODI, 2008). This was further endorsed by more donors under the Accra Agenda for Action in 2008.

The shift in emphasis from individual projects to SWAps shows a mixed record, however. In the case of agriculture, which has many parallels to climate change from the point of view of government service provision, SWAps have contributed to introducing more streamlined dialogues between donors and government, as well as strengthening government leadership and coordination across donors. This has, however, often happened at the cost of a re-centralization of policy making, particularly around central departments in the Ministry of Agriculture, and of an excessive focus on the SWAp process itself, rather than on sector policy outcomes. There is also very limited evidence that SWAps have actually led to a reduction in transaction costs – in fact, heavy management structures have been created to support the design, implementation and monitoring of the SWAp. Furthermore, and in spite of the original intention, agricultural SWAps have often focused on how resources are channelled to ministries of agriculture, but do little to stimulate linkages with other sectoral ministries and the private sector (Evans *et al.*, 2007).

Cabral and Shepherd (2008) argue that the aid practices that had dominated the development scene since the implementation of the structural adjustment programmes that were introduced with the globalization of liberalism did not foster ownership on the part of the recipient governments or countries. 'National "ownership" of reform processes is now widely thought to be critical to making sustainable progress on development and poverty reduction' (Cabral and Shepherd, 2008). This also led to a break with the then dominant aid practices.

When it comes to climate change adaptation, however, this chapter argues that it is not enough to foster ownership at the national level. Local governments are the most appropriate level for implementing national climate change adaptation and mitigation plans. This is because they are closer to the people and provide a natural interface between international investments, national policy frameworks and civil-society initiatives in addressing climate change issues at grassroots level. Especially with regard to spatial variations in climate change hazards from community to community, this chapter is inspired by the above quote from Cabral and Shepherd and argues that the local ownership of decision-making and reform processes is critical in coming up with sustainable institutional solutions and developments regarding climate change adaptation. When it comes to project implementation, however, funding is not devolved from the national to the local level, thereby disempowering the local authorities of their implementing power.

As in many African countries during the 1990s, Uganda adopted decentralization as the main form of governance of service provision at the subnational administrative levels. The 1997 decentralization act devolved powers in decision making and implementation to local governments at the district and sub-county levels. Decentralization is both a political process and a technical process. Political decentralization entails devolving leadership, decision making, participation, representation and relations with other governance structures down from the central government. Technical decentralization, similarly, involves the devolution of administrative, budgetary, planning, supervision, monitoring and evaluation aspects of programmes to the lower units of government.

Governance in itself is about the manner in which power and resources are used in the realization of development objectives. Good governance entails the efficient and effective use of the power and resources needed to bring about desired developmental outcomes. It is facilitated by constitutionalism, the rule of law, justice, electoral and participatory democracy, and popular participation in social and economic processes, among other things. Good local governance brings about desired transformations for the benefit of the people. The Climate Change and Rural Institutions (CCRI) studies (Friis-Hansen *et al.*, 2013) explored the ways in which local governments in Uganda responded to climate change-induced pressures in a decentralized resource-constrained environment with an expanding population.

Given the many challenges they face, local governments are, however, usually branded as inefficient, lacking in capacity and being corrupt or simply unwilling to support national programmes. Yet the problem seems to lie in the resources provided for local governance. The experience of this study indicates that, whereas local governments are the focal points for the implementation of climate change actions, they are not given adequate resources to match their mandates (Friis-Hansen *et al.*, 2013).

Chapter 2 of this volume carried out a subsidiarity analysis of rural climate change adaptation and concluded that the appropriate levels of decision making for rural climate change adaptation actions are the district and subdistrict (ward, village) because they are ideally situated to create an enabling environment that allows rural citizens to adapt to increased climate variability. This chapter also identified the importance of political ownership at both district and sub-country levels, as climate change adaptation is a long-term, continuous activity that requires the support of legitimate local institutions.

The Climate Change Context in Uganda

In Uganda the effects of climate change are quite evident, and recent studies have revealed a very bleak future for water resources, agriculture and the economy in general (USAID, 2013). The impacts of climate change are illustrated by the increased incidences of malaria in the colder highlands of Kigezi due to global warming, the melting of the ice caps on the Rwenzori Mountains and the increased occurrence and severity of extreme weather events like floods, droughts and landslides, as well as damage to all sectors of the economy. The effects of climate change are increasingly evident all over the country, but most especially in the highlands of Elgon, Rwenzori, Bundibugyo and the southwestern Kigezi highlands, and in the low-lying plains of the Teso and Lango subregions. These areas have witnessed very heavy rains, mudslides and flash floods, leading to deaths and the displacement of the population. Often the torrential rains and floods are followed by prolonged droughts or the delayed onset of the next rainy season, resulting in total crop failures and famine. These extreme effects on the population are often beyond what the local communities and government structures can handle, thus justifying the need for external science-based assistance to supplement local grassroots adaptation efforts and necessitating interventions like TACC.

Recent studies in Uganda predict up to a 40% fall in the production of staple food crops like cassava, potato and sweet potatoes, as well as export crops like coffee and tea. In recent years, climate-induced disasters have exerted enormous pressure on those governance structures that are closest to local people, yet they lack both the resources and the skilled human capacity to meet the present challenge (MWE, 2015). Nationally, the policy framework has a number of disaster management structures, including a fully-fledged Ministry for Relief and Disaster Management. Also, environment management and disaster reduction committee structures exist right down to the parish and village levels.

Overview of the TACC Project

TACC was implemented in three districts southwest of Mount Elgon Park in the context of the Bududa landslides and the excessive rainfall in the region, which affected livelihoods, food security and children's schooling. The Mt Elgon region experiences high population pressure on land, leading to the fragmentation of holdings, over-cultivation and encroachment on natural resources high up the mountains.

In the Mt Elgon region, the TACC project was implemented by UNDP with funds from the Danish Embassy, the UK Department for International Development (DfID) and UNDP, as well as technical and development support provided by the Welsh Assembly Government. TACC was a pilot project for the Global Initiative, 'Down to Earth: Territorial Approach to Climate Change', which sought to identify and develop projects that met local needs, while building climate resilience and the infrastructure needed for low-carbon growth.

Findings

Institutional mechanisms for the integration of TACC into local government

The TACC project in Bududa, Manafwa and Mbale Districts was implemented through two components over a 2-year period (2012–2013) with a total of US$1.5 million. The first component (90% of the budget) consisted of developing a Territorial Action Plan, whereas the second component (10% of the budget) involved a small grants programme (financed by the Global Environment Facility (GEF) and therefore aimed at mitigating activities rather than focusing on adaptation) aimed at involving local stakeholders. Seventeen out of 350 proposals received were funded by the TACC project to the tune of US$5000 each, two in Bududa, nine in Manafwa and six in Mbale.

The TACC project signed a Memorandum of Understanding with the three district administrations and all stakeholders were regularly consulted during the project period. However, all major decisions were taken by the project management in consultation with the Ministry of Water and Environment and UNDP. TACC activities were implemented through structures established by TACC in parallel with but outside existing local government structures. These included: (i) a regional project forum, which brought together all stakeholders interested in climate change and environmental protection, including non-governmental organizations (NGOs), the private sector and local government; (ii) establishment of a TACC project website (UNDP, no date); (iii) setting up a steering committee comprising TACC staff along with technical staff from the three districts; and (iv) establishing a Board of Trustees consisting of local leaders including Chief Administrative Officers (CAOs) and elected political leaders from the three districts. These were responsible for vetting proposals (component two) and commenting on project reports and decisions.

The TACC decision and implementation approach represents an instrumental and populist participatory approach, where local government leaders, staff and local stakeholders are consulted on all issues but not empowered to be involved as partners in prioritizing how the plan (component one) should be designed or how the available funding should be used. The invited political space given to district leaders was limited to prioritizing which NGOs should be supported on the basis of the criteria laid down by the project.

An officer within each district was assigned to be the TACC focal point within the district. This institutional structure is not compatible with the existing local government administrative structures. The salary for the focal point officer was (well) paid by TACC, to whom the officer reported. Meanwhile, the institutional and technical capacity to address climate change adaptation within the three districts, such as the Natural Resource Departments, was reported to be understaffed owing to the prevailing government ban of staff recruitment. The three local governments also lacked adequate extension staff because of the restructuring and centralization of the National Agricultural Advisory Services (NAADS), and therefore they had very few technical frontline staff who could provide services to the rural population on a day-to-day basis. In addition, the frontline staff lacked basic transport that would allow them to reach farmers. Very little or no financial support was provided by TACC to upgrade the capacity of local governments so that they could address climate change challenges using their own structures.

For the formulation of the Territorial Action Plan (component one), TACC chose to employ external consultants instead of involving local government staff. Although the formulation process involved extensive consultation among all stakeholders, local government political leaders and technical staff were not directly involved in the actual formulation process, including decisions concerning future implementation modalities. Some 70% of the TACC budget of US$1.5 million went on TACC staff, external consultants and overheads. This way of spending the available financial resources is perhaps the clearest indicator that the influence of the three local governments on the TACC project was very limited. As discussed in the next section, TACC's decision to use more than two thirds of the budget on TACC staff and consultants, instead of using it to take action directly, or at least to build local capacity to address climate change hazards, generated considerable resentment among local government staff and political leaders.

The TACC project treated the issue of integration with local government structures as an 'exit strategy' on its website (UNDP, no date), where it is stated that: '(1) strengthening linkage between Mbale region and National level is critical for sustainability of project results; (2) mobilizing resources for implementation of Integrated Territorial Climate Change Plan (ITCP) for Mbale Region including the investment package; and (3) the lessons learned and best practices to be disseminated and replicated to other parts of Uganda targeting mainly the hilly and mountainous areas'.

Forging links with local government is not an issue that can be fixed, however, by treating it as an add-on at the end of the project period. Whereas local participation was enlisted through a number of stakeholder platforms such as the ITCP steering committee, and the Mbale Region Climate Change Forum (RCCF) and website, these were parallel structures designed to legitimize the project, rather than representing any real devolution of decision making and capacity building within local governments. As for mobilizing resources to finance the ITCP, the districts have attempted this in the past 2 years, though without any success in attracting donor or government finance. The explanation given by the technical staff involved was that the plan is too expensive and that it is not a cost-effective way of addressing climate change adaptation. However, the key informers interviewed expressed ambivalence by stating that, if a donor were ultimately willing to finance the ITCP for an extended period and if local government were to be allowed a greater say during its implementation, it would certainly be highly popular among local government staff and leaders. As for the third element in the TACC exit strategy, the project is portrayed as a success story on the global UNDP/TACC website.

Political and technical inclusiveness of the planning and implementation of TACC

The ITCP is an investment plan for the period 2014 to 2019 and constitutes the main output of the TACC project. It was developed using a *planned climate change adaptation approach.* In its original form, this approach takes point of departure in climate science and is based on scientific expert knowledge. TACC combines this approach with extensive consultations among stakeholders, a process facilitated by highly trained international and national external consultants. The inclusiveness of the formulation process is limited to instrumental participation, a situation in which the intended beneficiaries are consulted but not directly involved in formulating the content, setting priorities for the budget or influencing the implementation modalities.

The TACC project website describes the process of formulating the ITCP as follows:

> TACC–Uganda project assisted the region to prepare for low carbon and climate change resilient developments and contributed to improving the livelihoods of people in the region to mitigate and adapt to climate change and poverty reduction. Specifically, the project supported local decision makers and planners to design and prepare regional integrated climate change adaptation and mitigation policies, strategies, plans and also formulated concrete actions and investment plans to be used for promoting long-term sustainability and poverty reduction in the region.
>
> (UNDP, no date)

Political and technical inclusiveness was much higher in component two, the GEF small grants window taking up 10% of the budget. This small competitive grants scheme was widely perceived as the most sustainable output of the TACC project by key informants interviewed during this study. As one informant put it: 'The process of developing project proposals from the local communities was inclusive, and it was the one which promoted ownership among the communities'. The flip-side of these activities is, however, that they are focused on climate-change mitigation rather than adaptation. Their main objective is thus to reduce emissions, with planting trees being by far the most popular activity.

Perception and sense of ownership of TACC among district officials

During a group discussion on the TACC projects, one district official remarked: 'TACC was interpreted by the political leaders as a (play)ground for consultants to "eat" free money, and this was a major concern among local leaders at the districts'. This sums up many statements from stakeholders about the TACC project. Many felt deep resentment over how the available funds were spent. As one elected local government politician expressed it: 'TACC was a project through which external consultants could siphon away free money meant for assisting local governments' response to climate change challenges'.

The combination of local government staff's and politicians' perceptions that: (i) they were not being taken seriously as political and technical partners; (ii) most of the funding was being spent on activities that were viewed as wasteful (external consultants); and (iii) implementation was being carried out outside existing local government structures, have meant that none of the informants who were interviewed felt they had ownership of the ITCP.

On the positive side, informants agreed that TACC had contributed to mass sensitization of the population to climate change adaptation measures and had initiated widespread tree planting. One environmental officer had a positive view of the TACC project, stating that: 'In fact, TACC can be rated as one of the most successful projects we have had. The ITCP plan was completed in time; nearly two-fifths of the people in the region were sensitized about climate change issues; and the small grants schemes to improve livelihoods are on-going such as biogas and agro-forestry farming'. Also, the local CBOs that received funding through the competitive small grants scheme were largely positive towards TACC, and many continued their tree-seedling activities on a commercial basis after the TACC project had ended. Another District Environment Officer said: 'the small competitive grant scheme helped to improve local livelihoods through start-up of community-based projects like biogas and agro-forestry tree planting'.

Several informants from local government agreed to discuss how things should have been done differently. One environmental officer noted that, if the TACC project had been implemented directly by the district structures, it would have created a good basis for climate change mitigation and adaptation in the region. This would have minimized the role of consultants undertaking scoping studies and directed their efforts more to capacity building within local government departments to enable them to make their own plans involving communities and to draw up more realistic cost estimates that could be included in national budgets. The district plans would then be merged to form the regional ITCP plan, in which every local government has a stake, and further support can also be given to implementation by means of a conditional grant from the Ministry of Local Government. Such grants are already being used to benefit other sectors, such as agriculture, health and education.

The institutional sustainability of TACC-initiated activities

At the end of the TACC project, an exit strategy was executed in an attempt to ensure the continued implementation of project activities. This involved establishing a new institution outside local government structures called the *Stakeholder Consultative Forum* that would be convened regularly by the Ministry of Water and Environment and would involve all local governments and NGOs active in the three districts. Interviews with informants during this study carried out 2 years after project completion indicated that the new Stakeholder Consultative Forum had only met once: 'Over a period of one year since TACC wound-up, only one stakeholders' meeting has been held to discuss how to raise funds for climate change adaptation. The meeting resolved to market the ITCP to donor agencies like the USAID, IUCN [International Union for Conservation of Nature] and the Ministry of Water and Environment but none of this has succeeded so far' (personal communication by a District Environmental Officer).

The three local governments of Bududa, Manafwa and Mbale were also requested to mainstream the plan into their district development plans. Because of the low sense of ownership of ITCP among the local governments and the fact that the plan is not an outcome of their own political process, very little was done after the TACC project ended. The ITCP has been shelved as a record of another project having been hosted and it has not been integrated into any of the three District Development Plans. The high turnover of local government staff undermines institutional memory and most of those who had been directly involved in the implementation of the TACC project had been transferred to other areas by the time of our study. At times it was a challenge for the study to identify local government technical staff who actually knew about ITCP and its intended purpose.

It is the small grants window that has had the most lasting effect. Two NGOs that partnered with TACC remained active. One is the ECO Trust, which continues to train farmers in the use of improved charcoal stoves, in how to access carbon financing and in coffee and banana production in all the three districts. The other is the Mbale Coalition Against Poverty (CAP), which has continued its work planting trees, with up to 10 million trees having been planted in all the three districts. These NGOs work in close collaboration with the Climate Change Department in the Ministry of Water and Environment.

During the implementation of TACC, 350 proposals for climate change adaptation were solicited from local CBOs. Of these, 20 were each awarded US$5000 and 17 were still active at the time of the study. These included six CBOs in Mbale District engaged in eucalyptus tree planting, charcoal briquettes, coffee-seedling propagation, agro-forestry and honey production; two CBOs in Bududa District involved in agro-forestry and improved cooking stoves; and nine CBOs in Manafwa District involved in tree planting, energy saving, heifer cows and biogas, agro-forestry and coffee growing, fruit tree cultivation, nutrition for HIV positive farmers, rehabilitation of degraded hills, soil and water conservation, and 'functional adult literacy'.

Conclusions

During the study, we examined how central government and international development agencies engage with subnational government structures. In the conclusion, we will reflect on the experience of implementing TACC projects in these three districts in eastern Uganda and ask whether this approach should be scaled up.

Assessed in terms of its own development logic, the TACC project was successful: it was administered efficiently and met its own goals, the ITCP was completed in time, 16 NGO/CBO projects were financed and implemented, and the majority of people were sensitized to climate change hazards. According to the project website, the experience is ready to be scaled up. The implementation modalities of the TACC project are not unique and are more likely to be closer to the mainstream implementation approach for internationally funded climate change projects.

This study has applied a broader analytical lens by examining the extent to which the implemented project was linked to subnational administrative and political structures. The findings reveal very serious reservations over the TACC project as a model to be scaled up in Uganda. The study's conclusions boil down to four aspects:

- Political aspects. The TACC project started by signing a Memorandum of Understanding with the three districts. However, during implementation of TACC the political leaders were effectively disempowered, and project activities were implemented using project structures. Although the leaders were consulted by TACC staff and consultants and had some influence on the choice of NGO/CBO projects through their membership of the TACC Board of Trustees, they had little or no influence on final decisions concerning the allocation of financial resources or the formulation of the ITCP. The study clearly shows that bypassing the local political establishment has a strong negative influence on the same politicians' sense of ownership of the project and its outcome (the ITCP). The direct consequence of this is that the ITCP has not been integrated into the District Development Plans for the three districts.
- Economic aspects. The skewed priority for fiscal resources in allocating the bulk of the budget to project staff, external consultants and overheads for the Ministry and UNDP, with some finance being provided for competitive NGO/CBO projects and very little to local government, clearly does little to create close links. The differences in access to resources between the project and local government is particularly aggravating in a situation in which technical positions are not filled because of a brake on employment, while the annual environmental budget of the sub-county is as low as a few hundred US dollars. Animosities aside, however, there is little doubt that, in spite of capacity problems, the local government system is likely to be many times more cost-effective than project implementation.
- Administrative aspects. Because of TACC's use of external consultants and its deploying a TACC staff member as the focal point in each local government administration, the project downgraded the role played by local government technical staff in formulating the ITCP from contributing partners to passive agents. Interviews with technical staff revealed that very little capacity development took place during TACC implementation. One technical officer told the study team that he felt like an invited guest when he participated in the TACC consultative forum. The study found that the institutional memory of TACC among local government technical staff was surprisingly low, which was fuelled further by the extensive relocation of staff to districts elsewhere in Uganda.
- Content of ITCP. The study did not analyse the relevance of the ITCP to climate change adaptation. Nevertheless it is fair to observe that the choice of approach to developing a plan for climate change adaptation is reflected in its focus and the types of solution chosen. The TACC approach is one in which external consultants interpret the aspirations of rural people affected by climate change hazards and combine this information with scientific knowledge about climate change, technical solutions, etc. This approach is likely to have a bias towards expensive technical and infrastructural solutions that required expert and consultant knowledge.

Aside from their high cost, the fundamental problem associated with such technical solutions is that they are presented as apolitical solutions. However, climate change adaptation in rural areas is intrinsically a political process, as several chapters in this book illustrate. Climate change adaptation that requires groups of local farmers to change unsustainable land-use management practices is associated with simultaneous social conflicts and reconciliation processes that require the support of local leaders and a socially inclusive planning process if adaptation is to be successfully implemented.

References

Adger, W.N., Agrawala S. and Mirza, M. (2007) Assessment of adaptation practices, options, constraints and capacity. In: IPCC (ed.) *Climate Change 2007: Climate Change Impacts, Adaptation, and Vulnerability*. Cambridge University Press, Cambridge UK.

Cabral, L. and Shepherd, A. (2008) Aid approaches and strategies for reaching the poorest. *Working Paper*. Overseas Development Institute, London.

Evans, A., Cabral, L., Wiggins, S., Greeley, M. and Kaur, N. (2007) *Formulating and Implementing Sector-wide Approaches in Agriculture and Rural Development: A Synthesis Report*. Global Donor Platform for Rural Development, Bonn, Germany. Available at: https://www.odi.org/sites/odi.org.uk/files/odi-assets/publications-opinion-files/3926.pdf (accessed 9 January 2017).

Friis-Hansen, E., Bashaasha, B. and Aben, C. (2013) Decentralization and implementation of climate change policy in Uganda. *DIIS Working Paper 2013:17*. Danish Institute for International Studies, Copenhagen.

Huppi, M. (2006) *Annual Review of Development Effectiveness 2006: Getting Results*. World Bank, Washington DC.

MWE (The Ministry of Water, Lands and Environment) (2015) Uganda National Climate Change Policy. Available at: http://ccd.go.ug/index.php/resources-publications/220-uganda-national-climate-change-policy (accessed 22 May 2016).

ODI (Overseas Development Institute) (2008) Accra 2008: the bumpy road to aid effectiveness in agriculture. *Natural Resource Perspectives* 114, ODI.

UNDP ([no date]) Climate Change Adaptation in the Mbale region of Uganda (TACC), Project Overview. Available at: http://adaptation-undp.org/projects/dc-uganda-tacc (accessed 15 September 2016).

UNDP (2013) *Analysis of Adaptation and Mitigation Options: Territorial Approach to Climate Change in the Mbale Region of Uganda Project*. Available at: http://adaptation-undp.org/sites/default/files/downloads/tacc_mbale_cc_adapation_and_mitigation_option_and_opps_report_final.pdf (accessed 11 April 2017).

USAID (2013) Uganda Climate Change Vulnerability Assessment Report. Available at: https://www.climatelinks.org/resources/uganda-climate-change-vulnerability-assessment-report (accessed 9 January 2017).

6 Creating Political Space for Climate Change Adaptation in Northern Ghana

Julie Fogt Rasmussen and Esbern Friis-Hansen

NGOs as Facilitators of Climate Change Adaptation

In recent years, growing attention has been paid to adaptation to climate change, including by international and national non-governmental organizations (NGOs). Many of the new sustainable development goals stress the urgency of combatting climate change and conserving the environment. In addition to these focuses, number 16 of the sustainable development goals addresses the issue of peace, justice and strong institutions. Subtargets for goal 16 are to 'develop effective, accountable and transparent institutions at all levels' and 'ensure responsive, inclusive, participatory and representative decision-making at all levels' (UN, no date). But how is this best done? This chapter explores the role that NGOs can facilitate in creating links between policy makers, local institutions and those affected by climate change by strengthening participation through a rights-based approach.

The concept of participation has been widely acknowledged within the development discourse through an increased focus on decentralization and community involvement. In 1994 the World Bank defined participation as 'a process through which stakeholders influence and share control over development initiatives, and the decisions and resources which affect them' (World Bank, 1994). According to Gaventa and Valderrama (1999), this definition of participation and stakeholders was seen as a means to encourage action and funding outside the public sphere and as such was not related to broader issues of politics and governance. Although this World Bank definition is more than 20 years old, the criticism posed by Gaventa and Valderrama (1999) is still highly applicable to the case of climate change adaptation. Most climate change adaptation policies are formed at high political levels where blueprints are designed based on scientific data. Little attention is given to indigenous local knowledge in those places where climate change is happening. There is a need to localize the decision making to these places and question who the stakeholders actually are in the case of climate change adaptation, an issue that relates to broader issues of politics and governance. This chapter argues that the political space concerning climate change adaptation needs to be reconfigured to create more adaptive government practices through political decentralization and participation. Currently, a re-politicizing of climate change adaptation through development is occurring, with more attention being given to NGOs. It is likely that during the next few years more funding for climate change adaptation will be channelled through NGOs. This chapter focuses on the NGO CARE International as an example, with a special focus on CARE's Tamale office in Ghana. CARE International seeks to work through a participatory rights-based approach to climate change adaptation in order to ensure representative decision making at all levels, from the local to the global.

On the basis of fieldwork in Ghana during the summer of 2015, this chapter explores the role of CARE Tamale through its Adaptation Learning Programme (ALP), implemented since 2010, and shows how the programme has created a new local political space for adapting to climate change, with a special focus on involving both community and local government in the process. Besides Ghana, the Adaptation Learning Programme (ALP) for Africa was implemented in Kenya, Mozambique and Niger. In Ghana the programme was implemented in the two districts of East Mamprusi and Garu-Tempane, located in the Northern and Upper East Regions of Ghana, respectively.

Integrating Participation into Cross-scale and Multi-level Governance

Localizing decision making to places where climate change takes place needs to enable the affected stakeholders to participate in the process of adaptation. Cornwall (2002a) argues that efforts to enhance participation in development have to reconfigure the space between citizens and the institutions that are linked to their livelihoods (this chapter only deals with state institutions). For this new participatory development, the *positioning* of actors becomes central. 'Participation comes to mean more than taking up invitations to participate, extending to autonomous forms of action through which citizens create their own opportunities and terms for engagement' (Cornwall, 2002a). This implies distinguishing between the invitation to participate and a claim to participate. Using these concepts it can be questioned what it actually means to participate: is participation activated simply by accepting an invitation to participate or does participation only happen through autonomous claims to participate? Within development there has been a focus on enabling participation through institutional frameworks and mechanisms for public involvement. Less attention has been given to participation in practice and how it actually addresses those at whom the participatory approaches are aimed. More approaches need to locate *spaces* for participation in the *places* where participation occurs to strengthen the claim to participate. This implies that political spaces are not a given but something that can be reshaped and created (Cornwall, 2002a).

While this chapter mainly focuses on creating spaces at the local level, climate change is a problem that spans multiple scales from the local to the global spheres, making it a challenge across scale. 'Cross-scale' is in this case defined as a spatial dimension in which climate change occurs (Cash *et al.*, 2006). It therefore becomes problematic when defining the consequences of climate change to being only *global, national* or *local* because they are one and all at the same time. When reducing climate change to one of the spatial scales, there is a concern that it will overlook effective institutional solutions, which is often done by governments to make the problems tractable (Cash *et al.*, 2006). The solutions must therefore also be addressed by several actors at all levels through multi-level governance at all affected scales because the solutions needed may vary from the local to the national and international levels. Multi-level governance of climate change adaptation incorporates both vertical and horizontal dimensions of governance by creating linkages between different actors. In this study vertical governance entails interaction between different levels of actors, for example, local, regional and national within Ghana. Horizontal governance characterizes the interaction between actors at the same level, in this case local government within the two studied districts in northern Ghana. Thus multi-level governance also takes place within different scales. In particular, vertical multi-level governance is of special concern for climate change adaptation.

Cash *et al.* (2006) argue that there is a mismatch concerning the matching of scale between local and indigenous knowledge and the level of where decision making takes place. 'This is seen in the dual problems of large-scale scientific knowledge that has little relevance to local decision makers, e.g. global climate models that are at a resolution that is not useful to subnational decision making, or local, tacit, or indigenous knowledge that is not seen as credible by national or international actors' (Cash *et al.*, 2006). Due to the extremes of climate change, local knowledge is not sufficient by itself and needs to be addressed by experts. Geographical conditions result in consequences that affect scales differently, however, and local knowledge can be of the greatest importance to adaptation. The latter is slowly gaining recognition within the field of climate change adaptation, with growing attention being given to cross-scale and multi-level governance of climate change.

So how to address the complexity concerning the mismatch between where the knowledge for the decisions made is being produced and where it is actually needed? Focus on decentralization as a tool for environmental and development policy has emerged during the past three decades, where activation of local participation is seen as one of the core elements of political decentralization (Agrawal and Gupta, 2005; Bebelleh *et al.*, 2013).

Effective political decentralization followed by adequate fiscal decentralization and local participation becomes vital to climate change adaptation. Realizing that climate change affects each community differently owing to their different geographical, cultural and socio-economic characteristics, adaptive capacity (Brooks *et al.*, 2005; IPCC, 2007) is also starting to be mainstreamed into climate change adaptation programmes. Because communities have different adaptive capacities influenced by these different compositions and are affected differently by climate change, a participatory bottom-up approach is a way to include the needs of individual communities by integrating local indigenous knowledge with that of large-scale scientific knowledge to enhance adaptation.

The Adaptation Learning Programme implemented by CARE International in 2010 aims to increase the participation of local communities in local decision making regarding climate change adaptation. Taking its point of departure in the above theoretical framework, this chapter aims at unfolding the role that NGOs can play in reconfiguring this political space and creating new spaces by unfolding how participatory approaches are integrated in and received by local communities and local governance in the context of enabling sustainable multi-level and cross-scale governance structures for climate change adaptation.

Context Setting

In sub-Saharan Africa (SSA) there are many barriers to adapting to climate change. A systematic literature review carried out by Antwi-Agyei *et al.* (2014) on the barriers to climate change in north-east Ghana suggests that 'households are constrained by financial barriers, socio-cultural barriers, institutional barriers, technological barriers and a lack of information on climate change characteristics' (Antwi-Agyei *et al.*, 2014, p.297).

Like most countries around the world, Ghana is also experiencing climatic change impacting on the livelihoods of Ghana's poorer population in particular, who are involved in climate-sensitive sectors such as agriculture, which employs 57% of Ghana's labour force (Antwi-Agyei *et al.*, 2012). The economic growth that Ghana has experienced in recent decades has reduced the number of people living in poverty. In 1991 about half the population lived in poverty, whereas in 2012 this number was less than a quarter (World Bank, 2012). This improvement in poverty reduction has, however, mainly been seen in urban areas, with poverty still being endemic in the rural areas, especially in the northern regions (Antwi-Agyei *et al.*, 2012).

Many factors influence the vulnerability identified in the northern regions, such as higher poverty levels, drier land conditions, unstable rain patterns and increase in hazards, which affects the northern population's dependence on subsistence farming (Darko and Atazona, 2013). These factors, which are different from those found in other regions of Ghana, emphasize the importance of carrying out locally oriented adaptation to climate change. CARE International focuses on integrating local communities into climate change adaptation.

Decentralization in Ghana

Ghana has a solid political decentralized framework for implementing participation. Since the 1980s, Ghana has embarked on democratic decentralization, with the first local government elections in 1988 consolidating the role of local governance. The 1992 Constitution supported this democratic decentralization, enhancing and transferring power and functions to locally elected governments. Furthermore, the 1992 Constitution established that decentralization should include citizen participation at all levels of government through inclusion in decision-making processes. This consolidation of local governance was emphasized by the 1993 Local Government Act, which introduced decentralized development with authority vested in the local district assemblies. From this a bottom-up approach and participation can be identified, where local communities are involved in the identification of development needs. The National Development Planning System Act 480 of 1994 states that the district assemblies should conduct public hearings for both primary and secondary stakeholders on all proposed district development plans. This Act 480 is supported by the 2006 Guidelines for the preparation of District Medium Term Development Plans (Ministry of Local Government and Rural Development, 2010).

There are, however, a number of constraints on realizing devolution, of which continued central government and ruling party control of local authorities and a lack of genuine fiscal decentralization are the most significant. Local government institutions and decentralized departments also lack sufficient human and material resources to effectively promote development in local communities. Decentralized planning therefore co-exists with central government planning and budgeting. Decentralization by devolution also coexists within a system of partly centralized regional coordination.

Currently, various initiatives are being spearheaded to deepen devolution. The District Development Fund (DDF) initiative is an intergovernmental fiscal funding modality that aims to harmonize the various flows of funding and provide better incentives for the District Assembly to improve performance using a performance-based grant-allocation system. Over the past decade, civil-society organizations and development partners have supported and implemented numerous social accountability initiatives, including support for local governance and community participation.

CARE and its Rights-based Approach

A rights-based approach has been central to CARE's work since 1999, when it launched its Human Rights Initiative. Through ALP, CARE aims to integrate a rights-based approach, which it is now implementing as part of climate change adaptation. The objective of ALP is to build the capacity of vulnerable people in sub-Saharan Africa to adapt to the impacts of climate change by developing approaches to community adaptation (CARE, 2015a). In Ghana, the programme was implemented in the two districts of East Mamprusi and Garu-Tempane, located in the Northern and Upper East Regions of Ghana, respectively. In each district, four of the most vulnerable communities were selected to participate in the programme, encompassing around 27,700 citizens. The main activities are small-scale rain-fed farming and livestock rearing, with secondary activities that can be described as petty trading and agricultural processing.[1] Due to the achievements of the programme, it has been extended to mid-2017 with plans to expand to the Upper West Region.

Community participation is central to the programme, which, through training and the strengthening of advocacy, aims to create linkages between local institutions and the involved communities to strengthen adaptation. Through participatory rural appraisal (PRA) workshops, the communities were invited to develop Community Adaptation Action Plans (CAAPs) whereby each individual community acting together identified its own specific needs to address vulnerabilities within it. A key element of the CAAPs is to include the different groups within a community so as to ensure that dialogue between the groups is created and that the end result reflects all community members, and not only the most resourceful. This emphasizes gender relations, where it is important for the community to reflect on gender relations in the community and on who benefits from the CAAP, as well as ensuring that the CAAP promotes greater gender equality. After this, the facilitation team, consisting of CARE, partners and experts, conducted a feasibility screening to see which ideas and projects it would be possible to implement in the different communities. By including local communities, CARE gives them ownership through their involvement in the project (CARE, 2015a, 2015b).

In operationalizing the rights-based approach in ALP, integration into local government was essential, an important aspect therefore being to establish linkages between the community and local levels. The communities initiated the process of establishing better links between the two levels through community monitors. In the beginning of the programme, CARE selected a few people from each community to become the community monitors. The selected people were mainly farmers, with a few teachers as well, reflecting the main profession within the communities. Most of the community monitors were young men and women who were among the most active within the community and who demonstrated both the will and the interest to be a part of ALP and take on the responsibility. The community monitors' role is to facilitate communication between local communities and local government. CARE provided extensive training to enhance the advocacy and communicative skills of the community monitors. The training of community monitors also ensures that each community has a spokesperson who can talk on behalf of the community and claim its rights.

Acknowledging that it is not sufficient to address only one of the many constraints of climate change, a more holistic approach is needed. It is possible to identify several strategies in ALP to reduce the constraints identified by Antwi-Agyei *et al.* (2014). Although this is not the purpose of this chapter, several strategies and elements of the ALP and how they aim at reducing these constraints will be presented throughout the findings with the aim of discussing how the initiatives have created a new political space.

Enabling Political Space

Since the implementation of ALP, all participants in the study, from both communities and local government, state that the programme has created new and better platforms for dialogue both within and outside the communities. These new platforms of dialogue are both reshaping and creating a new political space because they are developing new forms of communication and encouraging actors to communicate with each other. This institutional reshaping, which CARE has facilitated in the two districts, has been a two-way process involving an *invitation to participate* and strengthening the ability to *claim to participate* in the districts. Elements of the ALP are targeted at different actors, whereby some elements solely invite the communities, while other elements invite community, local government and experts to participate.

Interviewed members of both the communities and local government highlight Participatory Scenario Planning (PSP) as one of the main elements that have created new links between the actors at the different levels. When conducting the PSP workshops, CARE invited actors from the communities, local government and meteorological services to participate, thus serving as a platform bringing together local and expert knowledge and addressing the mismatch by integrating this combined knowledge into seasonal weather forecasts, helping local actors develop more climate resilient plans and enhancing farmers' long-term planning of their crops.[2] These PSP advisories are disseminated through a newly established climate information centre in Taraganga (one of the ALP communities), where speakers are set up in the village: as the village attracts many from nearby communities owing to its market, the information is disseminated to a larger crowd. Besides establishing links between different actors, many community members state that through PSP they have learned that communities are better at addressing climate change than individuals, thus shifting the focus from individual to community adaptation. Both local government and communities report a newfound willingness within the communities to assist each other, as well as helping other communities outside ALP: 'by helping your neighbours, all benefit', as one community member stated. Besides the forecasts, the climate information centre also reruns radio shows about climate change and related issues. The PSP and the climate information centre thus also serve an educational purpose in creating awareness of climate change.

Within the community, CARE has focused on creating social and gender-inclusive participatory processes and discussions by ensuring that all members of the community who wished to participate were invited to the CAAP workshops. A key element of the CAAPs is to include different groups within a community so as to ensure that dialogue between the groups was created and that the end result reflected the views of all community members, and not only the most resourceful. The importance of the community engaging in reflections about gender relations within it was stressed, as was the issue of who benefits from the CAAP, as well as ensuring that the latter promotes greater equality between the genders.

In the communities too, women especially have taken a key interest in ALP, as can be seen through their participation in the Village Savings and Loan Associations (VSLA), which has been positively received by both communities and local government. As a result of the popularity of VSLA, several groups now exist within the communities, with almost every household being part of one. Besides the economic possibilities in these groups, 50% (mainly women) of the community members interviewed stated that VSLA has also become a place to share information about climate change, agriculture and PSP advisories. The VSLA groups are one of the elements of ALP that has been highly acknowledged and spread to many communities outside ALP, with help from the VSLA focal persons in the ALP communities. Local government institutions such as the Department of Agriculture also highlight the enhancements in communication enabled by the VSLA groups. The Department of Agriculture provides farmers with field-school training (a season-long group-based transformative adult education programme) in the districts, which they only target at farmers. In the training sessions, however, there will always be a farmer who is part of a VSLA group and the farmer takes the newly acquired knowledge from farmer field schools back to the village and shares the information.

Women as Lead Actors in Climate Change Adaptation

The focus on gender equality through ALP and the extended invitation to participate has led to a great interest from women in community meetings, which has emerged through their own engagement and initiative. Both the community monitors and local government ascribe to the fact that women have received many benefits through the ALP initiatives such as VSLA and are now starting to claim more equal rights between men and women through participation. 'And then the women will tell you, because of VSLA there is now a good relationship between their husbands. Because in the past, the husbands used to be more powerful because of their economic power. But through VSLA the women have money' (staff from CARE Tamale). However, gender roles still affect this higher turnout from women because it is mainly men who engage in farming activities in fields far away from the villages.

Participation in VSLA has enabled women to increase their savings and thus contribute to the household economy mainly by paying education and medical fees. This has increased the economic role of women in the household. All those interviewed stated that since the implementation of ALP women have been included in household decision making, due mainly to the economic factor mentioned above. 'ALP has improved women's access to land. Now men and women hold household meetings together where the women are listened to, as well as on a community level' (female community member). Many of the women interviewed, however, still report systematic disadvantages within the agricultural sector. One woman in Jawani explained that it was easier for men to acquire a tractor than women because men most often rented them and it was easier for men to talk to each other than for a woman to talk to a man. Therefore many women sent their children to Tamale to rent a tractor. Acknowledging that there are systematic disadvantages between men and women, some projects have been targeted directly at women because of their vulnerability. An example of this is dry-season gardening implemented in Kugri and Taraganga. The women take turns in maintaining the community garden, something that has proved to be a success in terms of food security. In both villages, water and land resources already existed so all the villages needed was a water pump to maintain the garden during the dry season. The women are able to sell the surplus production in local markets, which provides them with a small income.

Besides the dry-season gardening, the men have started giving some of their plots to women so they can engage in their own agricultural production. The women thus control small areas of land: 'women got access to land, and became more independent' (female community monitor). Access to land and new economic capital are changing decision making in households, where women are acquiring greater influence and participating more. 'In fact, even females are more proactive than their men' (Department of Agriculture). The Department of Agriculture in East Mamprusi states that they find that women become involved with their crops more than men do, as can be seen through their attendance at farmer field schools, where sometimes more women than men now participate. Women's participation, the Department of Agriculture argues, is due to the fact that household burdens still fall upon the women, who therefore need to be more proactive in order to ensure the running of the household as well as their new agricultural responsibilities. This argument was backed by a female participant who stated that, while men may help the women in the fields, the household is still a woman's task.

Building Capacity for Adaptation to Climate Change

An important element of ALP is to create links between communities and local institutions for purposes of climate change adaptation. The main link between these two actors has been through the community monitors. The community monitors reside within the community and therefore have local understanding, enabling them to represent the community. The community monitors are in charge of communication between the community and local government, thus ensuring that the community is heard by local government and that information from local government is passed on.

The community monitors underwent extensive advocacy and communication training in the beginning of ALP, to equip them in the best way possible. 'That is part of the training. Because, alongside the community adaptation action plans, the communities were trained on advocacy and lobbying to have those skills to demand from duty-bearers. And so they have been trained. We have core groups in each of the communities who have some skills in advocacy. And they are those who move to put the pressure on the assembly' (implementing partner). '...For now it is mostly the ALP communities that have that capacity, also they have been trained in advocacy, lobbying and those skills. They know their rights more than any other communities. So they knock at our door more than the others, especially the community monitors' (member of local government). The community monitors work alongside the assemblyman (an established position within the Ghanaian political structure) at local governance. The main job of the assemblyman is to represent the community at local government. The assemblyman is elected by the community, unlike the community monitors, who are selected by an outside actor, in this case CARE. From this, one might assume that the assemblyman would be the best person to represent the community but, according to the majority of the community members who were interviewed, the community monitors do a better job at representing them than the assemblymen. This view is also supported by a member from local government: 'Yes, some assemblymen are not very useful. But what I will say with the community monitors is that they will give more capacity to really function as a unit or monitors. So the idea of the community monitors can be targeted as useful at the community level, so at a larger scale all the unit committee members are community monitors, and then the necessary capacity is built there because it is already a standing structure there. You see, it is one of the recognized structures as far as the local governance'. Both positions are voluntary but many argue that the fact that the community monitors reside within the community gives them greater local awareness and connectedness with the community members. The assemblymen are from the villages, but have often moved to nearby bigger cities with their jobs. Often their new profession does not reflect the main profession of the villagers, who are mainly farmers. Thus many aspects affect the assemblymen's representation of the communities. In three out of the four ALP communities visited for this study, there was frustration that the assemblymen did not reside in the community and therefore did not have an adequate understanding of the communities' needs.

According to both the community members and local government, the community monitors have conducted great advocacy that has benefited the communities, this being closely linked to their residence in the community. 'They [community monitors] have own voice, they push more and so on. They will knock at your door and tell you this is what we want, so we have good relations. But sometimes the resources restrain their ideas, development plans. It is not like, whatever is in your action plan we can come together and build it. It is a continuous project' (member of local government from Garu-Tempane). Examples of this lobbyism include a health clinic and dams that have been created in ALP communities with support from the district assemblies but, as already noted, a lack of resources is restricting development.

The advocacy and lobbying have strengthened the dialogue and collaboration between community monitors and local government. The community monitor is an example of an actor claiming to participate on behalf of their communities, and local government accepts these claims by inviting them into the political space of climate change adaptation. All the community members interviewed were positive about the new level of collaboration with local government and feel included in the process. They all expressed frustrations, however, because of how the lack of funding was hindering further development, as reflected in the statements of members of local government. Nevertheless, community members identify a political will in local government to help the communities, but also realize that the slow development progress is due to a lack of funding.

Not only is the increased interaction between the actors reshaping the political space: the extensive training and capacity building of the community monitors led to them running, in almost all of the eight communities, for district elections on 1 September 2015, thus challenging the position of the assemblyman. The interviewed community monitors argued that the main reasons for running were frustrations with the assemblymen and that the assemblymen did not represent the communities well enough. 'The assemblyman will do what he likes, not what the communities like' (community monitor). Also, many monitors reported a newfound interest in politics after being part of ALP. None of the community monitors, however, were elected to local government despite their large support within the community. Possible explanations might be that it is very difficult to run as an individual candidate because party politics dominates the political landscape in Ghana. Also, community monitors are essentially farmers and not politicians. As one community monitor stated: 'The assemblyman tries to blame the community monitors when things go wrong'.

Institutionalizing Participation

In both the districts where ALP has been implemented, CAAPs have been integrated into the districts' Medium Term Development Plans for 2014–2017. The district assembly of Garu-Tempane has, on its own initiative, expanded the CAAPs from the four ALP communities to 98, with the aim of implementing CAAPs in the entire district with more than 200 communities. Currently, a lack of funding and staff is hindering this. The National Development Planning Commission has also included CAAPs in the District Planning Guidelines for Medium Term Development Plans. Focus on integrating multiple levels of stakeholders has been steady throughout the programme. CARE has extensively trained the national level in the benefits of community participation. ALP was launched at the same time that Ghana was drafting its National Climate Change and Adaptation Strategy and staff from CARE Tamale stated that through lobbying they were able to integrate community-based adaptation into the strategy. CARE facilitated workshops on the scope of ALP and community-based adaptation for National Development Planning Commission staff, who then travelled to Ghana's 216 districts and trained the district planning officers on the guidelines of CAAPs.

Along with VSLA and PSP, CAAPs are the programmes that have had the greatest spill-over effects into neighbouring communities. PSP was introduced to the Ministry of Food and Agriculture, which has recognized the benefits of the programme, is now collaborating with the Ghana Meteorological Agency and has incorporated PSP into its sectorial policy. The success of PSP has led to members of district assemblies from Togo participating in PSP workshops. PSP documents the positive outcomes that the integration of multiple levels of stakeholders can exert on climate change adaptation. PSP has expanded on the African continent and is currently being replicated in Ethiopia, Malawi and Tanzania (CARE, 2015c). This is also due to the fact that CARE International has worked on a global level. According to CARE's own evaluation report, globally ALP has achieved recognition for its approaches by enjoying good visibility at the International Institute for Environment and Development (IIED), while hosting international conferences on community-based adaptation and giving presentations at the United Nations Framework Convention on Climate Change (UNFCCC) Conference of Parties (CARE, 2015c). Figure 6.1, from CARE, illustrates the spread of ALP in Africa.

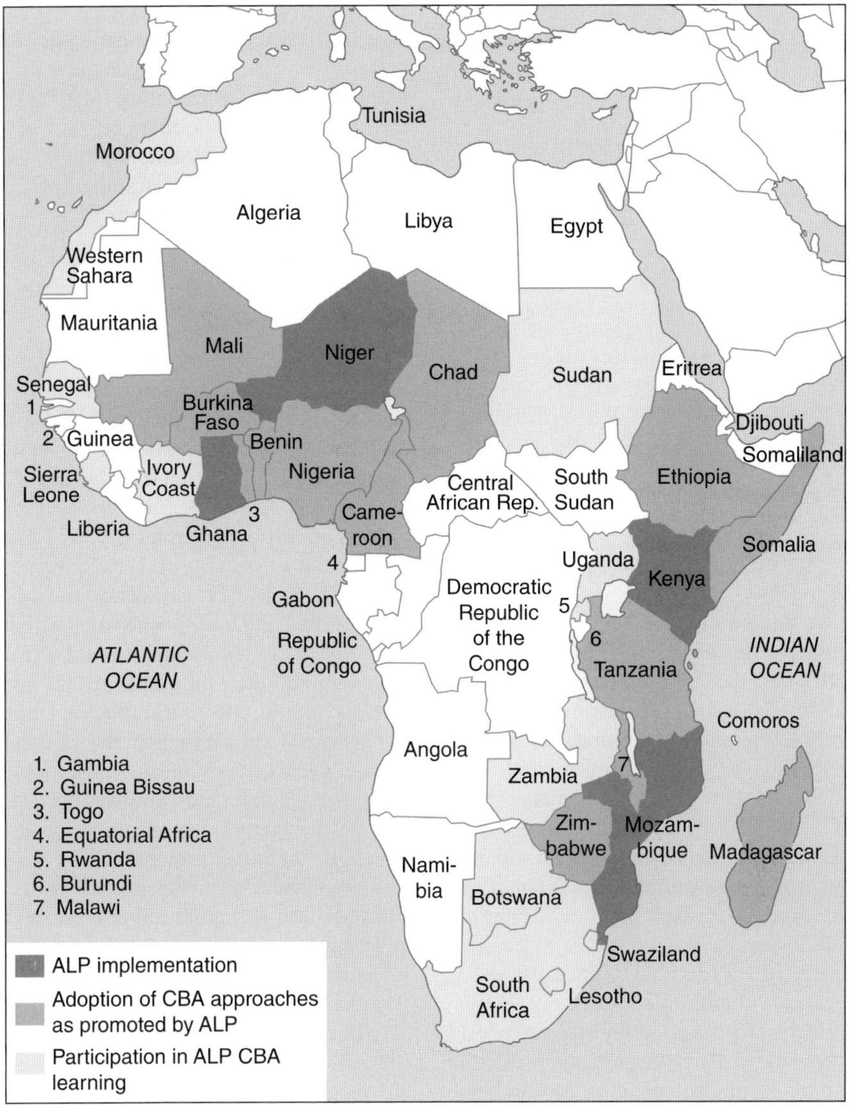

Fig. 6.1. Adaptation Learning Programme (ALP) reach in Africa 2015. CBA, community-based approach. Source: CARE International's *Adaptation Learning Programme for Africa* (CARE, 2015c).

In Ghana, the lack of funding within local government was cited as a major frustration by everyone who was interviewed. Fiscal decentralization is not being put into practice effectively, thus hindering further development. 'National level. . . . It is like for the last four years they have forgotten us' (Department of Agriculture). Communities recognize the political will of the district assembly and acknowledge that the slow progress is due to a lack of funding. Through its global engagement and lobbyism, among many others, CARE International has contributed to supporting the proposal that the Green Climate Fund, the financial mechanism under UNFCCC, should allocate 50% of its funds to climate change adaptation (CARE, 2015c). There is currently no mechanism, however, that allows resources from the Green Climate Fund to be allocated directly to local government in any given country.

The institutionalization of ALP elements in Ghana is a reflection of the legitimacy of the programme. As the following quote describes, there are many elements that feed into the positive outcomes seen in Ghana:

> 'And if you look at what comes in in term of finance and other support [from the national level], it is far less to deal with those things. So any organization that comes in in any form [that] tries to deal with issues such as climate change should be accepted. And really embed this. And we can see this really happen… But sometimes the education at the lower level also plays a very important role. But the district assembly cannot move to every community to give education. But when CARE came and they started giving training through ALP, you know there are some communities I can proudly say that you get educated… And these are some of the good things we are seeing. So clearly it has been good.'
>
> (Local implementing partner)

From this quote, one can infer that any actor entering the region with funding will be accepted, which can be ascribed to the fact that local funding is so low, this being identified by all those interviewed as the reason for the slow implementation or lack of implementation of adaptation and development initiatives. However, following one of the main arguments of this book, the question of adaptation is not only about having adequate funds but also about how the funds are used. CARE's practice of channelling funds to those affected by climate change at the local level is, as this book argues, the way to finance climate change adaptation.

Furthermore, the positive outcomes of ALP must also be viewed in their proper context. ALP has also been implemented in Mozambique, where it had to be abandoned due to difficulties with the project. The context therefore needs to be reviewed when measuring the outcomes of the programme in each country and considering how to transfer the programme to other countries. The political system in Ghana is very stable and in recent decades the country has embarked on a good development path. The outcomes must also be viewed by the facilitating team in each country, where CARE Tamale's facilitators are very actively engaged in ALP, while the staff who joined this fieldwork study were or had close relatives in the districts studied. From the outset, the close ties to the area and the familiarity with local languages created a good foundation for the project's implementation. Therefore the positive outcomes of ALP must be viewed holistically, with broader links to both the context of the specific locations and the actors involved. In these two districts in Ghana, all actors involved in enhancing the development achieved show a strong willingness to do so.

Participation as a Tool for Climate Change Adaptation

Rooted in the premises that climate change affects each community differently and that communities have different adaptive capacities, participation can be a vital tool in adapting to climate change. As Cornwall argues (2002a), there is a distinction between the *invitation to participate* and the *claim to participate*. For CARE, one of the key elements to ALP is creating links between communities and local governance, with the aim of minimizing the institutional barriers that many communities face. The focus is not, however, on communities to merely accept invitations to participate but is also on their claiming their right to participate in decision-making processes regarding climate change adaptation. Community participation is central to the programme, which, through training and the strengthening of advocacy, aims to create links between local institutions and the communities involved in order to strengthen adaptation. CARE grants communities ownership by involving them in the project (Ward and Percy, 2013; CARE, 2015b).

In the case of climate change adaptation, the invitation to participate in the political space entails that state institutions (in this case local government) invite the communities affected into the planning and decision-making process. This invitation to participate is not to be confused with the participatory invitations that CARE extends to both the communities and local government. On a theoretical level the invitation to participate offered by CARE and local government should not be narrowed into one single invitation, but instead represent different actors' offers of invitation into different steps of the process of climate change adaptation. The matter of who gives the invitation to participate is important to indicate if the project has become locally bounded. To ensure the sustainability and legitimacy of the project, state institutions need to take over the responsibility of inviting communities to participate that CARE so far has facilitated.

ALP has adopted a holistic approach to climate change adaptation. From the findings it is possible to show that ALP targets several of the constraints identified by Antwi-Agyei *et al.* (2014). VSLA and dry-season gardening address the financial barriers, as well as impacting on the socio-cultural barriers through the empowerment of women. In addition, the establishment of community monitors has led to better access to institutions, whereas PSP and farmer field schools have minimized the technical barriers while providing climate change information. Together, the many elements of ALP have a holistic reach that enhances adaptive capacity by addressing the underlying causes of vulnerability. This holistic understanding is also what has strengthened the individual's and community's ability to claim to participate.

In both the districts involved in this study, local governance has shown a great willingness to invite local participation, in addition to communities claiming their right to participate. This indicates that the project can sustain itself once CARE is no longer involved. Furthermore, the new links between the communities and local governance act to consolidate ALP's legitimacy and that of the new political spaces it has created. When working with participation, legitimacy is important to the process, since this increases participation further as well as the motivation for climate change adaptation, because communities can see the outcomes of their participation. In this study, it is clear that the invitation and the claim to participate have been integrated because both local government and the communities acknowledge the outcomes of ALP, for which no single actor is taking credit. Acknowledging that the outcome has been a collective process on the part of all the actors involved serves as a great argument as to why cross-scale and multi-level governance is so important for climate change adaptation. Participatory cross-scale and multi-level governance address this mismatch in decision making, which has characterized the majority of climate change adaptations in recent years, by including participatory practices.

The Limits to Participation

Since the turn of the century, there has been a critical international debate over how participatory approaches are used in development co-operation. Awareness has grown that social structures, cultural practices and local political contexts are not being adequately considered, resulting in existing hierarchies being reinforced, as well as continued social exclusion and economic marginalization. It has also become clear that attempting to empower poor groups without exacerbating relations with other groups is difficult (Mohan and Stokke, 2000).

As a reaction to contemporary criticisms of popular participation, Hickey and Mohan (2004) identify three tendencies in the use of participatory approaches: (i) widening impact and social inclusion; (ii) broadening agendas; and (iii) rejecting participation in its current form. ALP in Ghana is an example of the second tendency in that it broadens the agenda to capture a more political sense of agency and convergence between participatory development and governance. The assertion that development can be managed through the right mixtures of institutional responses has depoliticized development interventions (Ferguson, 1994) and transformed them into a technocratic process to be administered and planned for by agents of development. Cooke and Korthari (2002) have questioned the extent to which participatory approaches lead to transformative processes among poor people, whereas others take a more pragmatic approach by asking what is required to (re)establish the legitimacy of participation (Hickey and Mohan, 2004).

The activities of NGOs such as CARE and ALP can be seen as pragmatic attempts to make participation more transformative and empowering by shifting the focus beyond the individual and local to involve multi-scale strategies that encompass the institutional and structural aspects of governance. They do this by facilitating invited spaces that allow local communities a voice in policy-making processes and possibly influencing the agenda and rules of engagement. There are, however, limitations as to whether such invited political spaces foster critical consciousness and inclusive decision making, or whether they remain a methodology for improving project performance.

A critical analysis of different spaces for participation is therefore increasingly important (Brock *et al.*, 2001). In particular we need to pay attention to what is needed and to the fact that development is created by formal agencies of development flowing from us to them and, depending upon the knowledge, institutions and best practices of professionals (Cornwall, 2002b).

> Participation represents an authentic attempt to include others in decision making. Beyond the intent of the exercise, or even the amount of 'real influence' that participants get to have over decisions made, the problem of participation runs deeper. It is created by a particular way of thinking about social change, one that is deeply embedded in how formal development organizations think about themselves (as change agents) and about development (as something that they do).
>
> (Eversole, 2010)

Conclusion

This chapter has shown how international and national NGOs can facilitate the process of creating links between local communities and local government. ALP has centred around participation, with a focus on increasing invitations to participate while strengthening the claim to participate. In both districts studied it was found that the local communities and local government both engaged in the process of strengthening participation, thus making climate change adaptation a collective process. CARE has therefore facilitated this process of reshaping as well as creating a political space between local governance and local communities by inviting both actors into the process. The results achieved by ALP have led to spill-over effects into neighbouring communities, as well as some practices such as PSP that are being replicated in the whole African continent.

Because climate change affects each community differently, and because needs and vulnerabilities vary greatly even within very small geographical areas, acknowledging communities' claims to participate becomes vital to developing effective adaptation plans. Individual communities are best at identifying their needs and have different adaptive capacities, meaning that, conversely, applying best-practice solutions to several communities can result in lower results with the same or perhaps even higher costs. Therefore, more cross-scale governance with a special focus on the local scale can be influential in improving adaptive climate change governance if scale governance is effectively integrated with multi-level governance. As the case of northern Ghana shows, one way forward for climate change governance is to strengthen decentralization. Seeing that this is actually in place on a theoretical level, it needs to be accompanied by adequate fiscal decentralization. Both communities and local governance highlight the lack of funding as the main problem for further development. Therefore the structure of both the financing and governance of climate change adaptation needs to be revisited to determine whether or not both governance and finance can be decentralized even more to ensure better outcomes.

Notes

1 Numbers and descriptions from CARE Tamale.
2 For more about PSP, see Ward and Percy (2013).

References

Agrawal, A. and Gupta, K. (2005) Decentralization and participation: The governance of common pool resources in Nepal's Terai. *World Development* 33, 1101–1114.

Antwi-Agyei, P., Fraser, E.D.G., Dougill, A.J., Stringer, L.C. and Simelton, E. (2012) Mapping the vulnerability of crop production to drought in Ghana using rainfall, yield and socioeconomic data. *Applied Geography* 32, 324–334. Available at: http://dx.doi.org/10.1016/j.apgeog.2011.06.010 (accessed 12 January 2017).

Antwi-Agyei, P., Dougill, A.J. and Stringer, L.C. (2014) Barriers to climate change adaptation: evidence from northeast Ghana in the context of a systematic literature review. *Climate and Development* 5529, 1–13.

Bebelleh, F. Der, Nobabumah, A.S. and Region, U.W. (2013) Political decentralization and local participation in Ghana : perspectives from the Upper West Region. *Public Policy and Administration Research* 3, 12–26.

Brock, K., Cornwall, A. and Gaventa, J. (2001) Power, knowledge and political spaces in the framing of poverty policy. *IDS Working Paper 143*. Available at: http://www.ids.ac.uk/files/wp143.pdf (accessed 12 January 2017).

Brooks, N., Adger, W.N. and Kelly, P.M. (2005) The determinants of vulnerability and adaptive capacity at the national level and the implications for adaptation. *Global Environmental Change* 15,151–163.

CARE (2015a) Adaptation Learning Programme for Africa (ALP). Available at: http://www.careclimatechange.org/files/adaptation/ALP.pdf (accessed 12 January 2017).

CARE (2015b) Adaptation Planning with Communities. Available at: http://careclimatechange.org/wp-content/uploads/2014/08/CBA_Planning_Brief.pdf (accessed 12 January 2017).

CARE (2015c) Adaptation Learning Programme for Africa (ALP) – ALP Results, Outcomes and Impacts Report January 2010 to June 2015. Available at: http://careclimatechange.org/wp-content/uploads/2016/03/ALP-2010-15-Report.pdf (accessed 12 January 2017).

Cash, D.W., Adger, W., Berkes, F., Garden, P., Lebel, L., Olsson, P., Pritchard, L. and Young, O. (2006) Scale and cross-scale dynamics: governance and information in a multilevel world. *Ecology and Society* 11, 8. Available at: http://www.ecologyandsociety.org/vol11/iss2/art8/ (accessed 7 May 2016).

Cooke, B. and Korthari, U. (2002) *Participation: The New Tyranny?* Zed Books, London and New York.

Cornwall, A. (2002a) Locating citizen participation. *IDS Bulletin 33*, pp. i–x. Available at: http://doi.wiley.com/10.1111/j.1759-5436.2002.tb00016.x (accessed 12 January 2017).

Cornwall, A. (2002b) *Making Spaces, Changing Spaces: Situating Participation in Development*. IDS Working Paper 170, Institute of Development Studies, Brighton, UK.

Darko, E. and Atazona, L. (2013) *Literature Review of the Impact of Climate Change on Economic Development in Northern Ghana Opportunities and Activities*. Overseas Development Institute, London.

Eversole, R. (2010) Remaking participation: challenges for community development practice. *Community Development Journal* 47, 29–41.

Ferguson, J. (1994) *The Anti-Politics Machine: Development, Depolitization and Bureaucratic Power in Lesotho*. University of Minnesota Press, Minneapolis, MN.

Gaventa, J. and Valderrama, C. (1999) *Participation, Citizenship and Local Governance: Background Note Prepared for Workshop on 'Strengthening Participation in Local Governance'*. Institute of Development Studies, Brighton, UK.

Hickey, S. and Mohan, G. (eds) (2004) *Participation: From Tyranny to Transformation? Exploring New Approaches to Participation in Development*. Palgrave, New York.

IPCC (2007) *Climate Change 2007: Impacts, Adaptation and Vulnerability, Contribution of Working Group II to the Fourth Assessment Report 2007*. Cambridge University Press, Cambridge, UK.

Ministry of Local Government and Rural Development (2010) *Draft Decentralisation Policy Framework: Accelerating Decentralisation and Local Government for National Development*. Available at: https://www.giz.de/en/downloads/en-national-decentralization-policy.pdf (accessed 19 January 2017).

Mohan, G. and Stokke, K. (2000) Participatory development and empowerment: the dangers of localism. *Third World Quarterly* 21, 247–268.

UN (No Date) Sustainable Development Knowledge Platform. Available at https://sustainabledevelopment.un.org/sdg16 (accessed 7 May 2016).

Ward, N. and Percy, F. (2013) Community Based Adaptation: An Empowering Approach for Climate Resilient Development and Risk Reduction. Available at: http://www.careclimatechange.org/files/CBA_Brief_nov_13.pdf (accessed 19 January 2017).

World Bank (1994) The World Bank and Participation. Available at: http://www-wds.worldbank.org/external/default/WDSContentServer/WDSP/IB/1994/09/01/000009265_3961007180034/Rendered/PDF/multi_page.pdf (accessed 7 July 2016).

World Bank (2012) Country Info Ghana. Available at: http://data.worldbank.org/country/ghana (accessed 18 April 2016).

SUGESTI
BOX

7 Local Politics of Climate Change Adaptation in Uganda

Esbern Friis-Hansen and Charles Aben

Climate Change on the Political Agenda of Local Government

Climate change has a range of serious effects for the rural population in Africa. In recent years, East Africa has suffered from frequent episodes of both excessive and deficient rainfall. For Eastern Uganda, in particular, the frequency of anomalously strong rainfall causing floods has increased (Shongwe *et al.*, 2010). This is collaborated by qualitative interviews among the rural people of Teso who perceive floods to have become a significant factor that influence their livelihoods. The combination of gradual climate change that has distorted the traditional growing seasons and hazards associated with extreme climate change events such as floods or droughts have exerted pressure on district and subdistrict governance structures that are closest to the local people. Local governments in Uganda are well placed political institutions to examine the climate change hazards and act by creating an enabling environment allowing rural citizens to adapt (Friis-Hansen *et al.*, 2013).

In spite of the elaborate decentralization structure and regular elections of local government councillors, climate change is, however, rarely discussed as a political issue in the district council. Four years into implementation of the National Climate Change Policy (NCCP), central government funding for climate change adaptation is still absent in all of the District Development Plans in Teso Region. The absence of a budget line for climate change adaptation has contributed to local government politicians being inactive. Even though local government politicians at times come under pressure to act when they engage with their constituencies, the general trend is that without a budget line, climate change adaptation is not on the district council political agenda. 'No funding – no activities', as one local government politician put it during a meeting organized by the authors that brought together politicians from the nine districts affected by the floods in Teso and Karamoja Regions (Friis-Hansen *et al.*, 2013). Another politician at the same meeting hinted at another reason why local government politicians are not very articulate and outspoken when it comes to climate change adaptation. He said 'There are no votes in climate change adaptation'. Qualitative interviews with local government politicians and technical staff suggests that the climate change challenge is perceived as complex and very challenging to solve, because it involves adaptive natural resource management solutions that may be beneficial and lower risk for many people but are likely to aggravate others (many of which are among the local elite) as their economic activities may be limited.

This study explores the importance of political engagement and support from elected local government politicians for the implementation and institutional sustainability of climate change adaptation. Amuria district is deliberately selected for empirical fieldwork as a unique example of a local government in Eastern Uganda where climate change has been a contested issue in district council. During the 2006 election, an opposition candidate was elected chairman of Amuria district council on a platform that included addressing climate change adaptation. The study first explores how adaptation of climate change came to be on the political agenda of Amuria local government. With an aim to draw policy perspectives, the study seeks to understand how rural people's climate change concerns were articulated in a manner that translated into popular votes during election. The study further examines how a political mandate after election victory translated into support for climate change adaptation. Specifically, the study examines the experiences with participatory planning and implementing Wetland Management Plans (WMPs) in response to the extreme 2007 floods.

Empirical Context

Amuria is one of eight districts in Teso Region that is situated in Eastern Uganda. Although Teso Region historically has been subjected to occasional floods or droughts (i.e. 1972, 1997), the climate situation has significantly worsened since 2007, with erratic weather conditions becoming the norm. The region experienced widespread floods in 2007, 2009/2010 and 2012 that have devastated the region, destroying crops and food reserves, damaging roads, mud houses and other infrastructure, contaminating water sources and forcing many to temporarily resettle (ActAlliance, 2010). The 2007 flood was most widespread in Teso Region and the only one (out of the three floods) that received attention and support from central government and international disaster relief agencies (Epilo, 2011).

The perception among farmers and local authorities in Teso Region is that the rainfall pattern has gradually become unstable, reflected in statements such as 'rainfall has become more unpredictable' and 'the long and short rainy seasons have swapped places'. These perceptions are reflected in meteorological data. 'Normal' weather pattern used to consist of two distinct rainy seasons: the first season from March to June with a dry spell in July; and second rains running from August to November with a dry spell from December to February. Analysis of long-term precipitation data indicates that the total annual rainfall in Teso Region has increased; however, the onset and cessation of rains have become less certain. As a result, rainfall during the first rainy season has become unpredictable. When it rains, however, the rainfall is more intense and 'violent' (Bashaasha *et al.*, 2011).

When Teso Region experienced extreme floods in 2007, climate changes were neither widely understood nor appreciated by planners or politicians at district and national level and were therefore not an important priority in planning. There were few non-governmental organizations (NGOs) in the Teso region that focused on issues related to climate change, mainly in terms of tree planting (mitigation).

Responses by meso-level institutions in Teso Region to the 2007 flood were limited to helping with the delivery of emergency assistance, spearheaded by the United Nations High Commissioner for Refugees (UNHCR) and the World Food Programme (WFP). While legally performing the coordinating role, in practice the Department of Disaster Management and Refugees under the Office of the Prime Minister (OPM) contributed little in addition of permitting UN agencies and NGOs to carry out emergency relief. During an interview in the OPM it was made clear that the Ugandan government has allocated very limited resources to providing relief and rehabilitation support in response to emergencies. The OPM therefore largely relies on the UN emergency agencies, which only mobilize their resources during major emergencies. As a consequence communities that are affected by extreme climate events that may be equally severe, but more limited with number of people/area affected fall below the radar and remain without any central government support. Local government is under a considerable pressure to assist, but it does not receive funding from central government for such activities and its assistance therefore often remains symbolic.

There are three sets of reasons why the governance of emergency assistance policy is centralized and non-compliant with the Decentralization Act. Firstly, it is politically important to be seen to respond to emergencies and to ensure that assistance is seen to be attributed to central government. Secondly, control over when and when not to declare an emergency is important.

Historically, emergencies in Uganda have been closely linked to civil conflicts. Thirdly, but not least, the resettlement of displaced people in camps as a consequence of emergencies is politically highly sensitive because of its security aspects (Friis-Hansen *et al.*, 2013).

At the time of the 2007 flood, there was no comprehensive climate change policy and support for climate change adaptation was largely driven by externally funded project interventions implemented by central ministries with largely urban-based activities. The farmers were not aware of the hazard and the risk and it did not occur to them that there would be erratic weather changes that would result in flooding. Before the 2007 floods the farmers had no investments related to climate change and disaster risk reduction and there was little use of wetlands and unsustainable cutting of trees. The farmers mainly used their traditional farming methods with limited use of improved seeds and planting in rows. At the district level there was no disaster plan that would have been used for the response.

Analytical Framework

The study analytical frame combines two theoretical traditions, political ecology and institutional bricolage.

Political ecology

Political ecology allows us to understand the effects of policies and market conditions from local people's perspectives. Moreover, it is helpful to understand the political forces at work in the struggle for land-based resources and coping strategies at individual, household and community levels to resulting changes (Robbins, 2004). Society's interaction with land-based resources is linked to local, regional and global power structures. And 'power asymmetries and social inequalities shape patterns of land use and the management of natural resources' (Cleaver and de Koning, 2015). Political ecology is a useful approach to understanding the underlying processes of environmental problems such as environmental degradation and marginalization because it 'combines the concerns of ecology and a broadly defined political economy. Together this encompasses the constantly shifting dialectic between society and land-based resources and also within classes and groups within society itself' (Blaikie and Brookfield, 1987).

Governance and natural resource management users are referred to as actors. Actors are individuals and/or groups with active interest (users, direct or indirect) or vested interest (traders, NGOs, etc.). They are associated, but with different functions, powers and accountability relations that depend on the historical socio-political structures. Actors can be differentiated on the basis of their ideology, wealth, objectives, internal structure of their organization, membership, funding sources and the laws to which they are subjected. Characterizing actors is therefore important in understanding socio-ecological dynamics, especially in times when the interest of different levels of actors in pursuit for survival widely contrast.

Mainstreaming support for climate change adaptation

The United Nations development programme defines climate change as the development issues of our generation. The scientific community and development agencies agree that climate change triggers profound transformation of social and ecological systems that would cause significant dislocation and stress among affected populations (IPCC, 2007). Climate change adaptation has, during the past five years, been mainstreamed into a wide spectrum of national and international development agencies and NGOs. The dominant discourse of climate change adaptation relates closely to narratives of climate change produced by a few key international agencies such as IPCC, UNDP and the World Bank, who are producers of knowledge and strategies on climate change adaptation. The aim is to make existing development investments more resilient to climate variability and extreme weather pattern events while simultaneously improving the impact of development itself in the present (World Bank, 2006). The logic seems to be that, without climate change adaptation, the gains of development could be lost and, without development, societies remain vulnerable to the impact of climate change (IPCC, 2007).

Key concepts of mainstream climate change adaptation in relation to political ecology

This dominant narrative of planned climate change adaptation has three key concepts (vulnerability, resilience and adaptive capacity) that have been mainstreamed into developing thinking alongside good governance, empowerment and sustainability:

- Vulnerability: the degree to which the system is susceptible to, and unable to cope with, adverse effects of climate change, including climate variability and extremes. Vulnerability is a function of the character, magnitude and rate of climate change and variations to which the system is exposed, its sensitivity and its adaptive capacity.
- Adaptive capacity: the ability of the system to adjust to climate change (including climate variability and extremes) to moderate potential damage, to take advantage of opportunities to cope with the consequences of change.
- Resilience: the ability of the social ecological system to absorb disturbances while retaining the same basic structure and ways of functioning, the capacity for self-organization and the capacity to adapt naturally to stress and change (IPCC, 2007).

These three concepts are projected to be universally valid, in spite of the different character of challenges that each country faces and the diversity of socio-ecological settings within and across these countries. Adaptation has its roots in evolutionary biology, however, where organisms continuously adapt to changes in the external environment as a means to adjust to new constraints and opportunities.

The concept of *vulnerability* provides the conceptual bridge between the notion of the external climate and its impact upon society. Vulnerability is a concept that underpins the notion of adaptation as a process of reducing exposure to climate stress. Political ecology views vulnerability as a socially constructed situation that existed prior to the shock mediated by context-specific local social structures and institutions. Vulnerability 'is driven by social action that reinforces self-interests on the distribution of power in addition to interacting with physical and ecological systems' (Adger, 2006, p.279).

> 'The idea of *adaptive capacity* is produced as a new and universal object of development upon which the transformative practices of states, institutions and organizations can be set to work. The building of adaptive capacity, therefore, provides the universalizing glue that binds to cut the climate change adaptation as a core element of government practice' (Taylor, 2015, p.67).

Political ecology argues that adaptive capacity needs to be understood as a *relational term* in which ecological relations can mean security for some people, whereas it might mean insecurity for others. Political ecology offers a historic contextual analysis that allows a better understanding of conflicts between different land users.

The separation of society and climate underpinning the mainstream climate change approach

Three-quarters of the academic publications on climate change adaptation viewed adaptation as a technical process of planned social engineering to guard against proximate climate threats according to a literature review by Bassett and Fogelman (2013). This climate change adaptation narrative is based on the assumption of dichotomy between climate and society, which is seen to represent two distinct domains, one physical and natural and the other cultural and social, that relate to each other as separate entities. Adaptation emerges as the logical process of planned social adjustment to climatic threats.

However, by assuming a dichotomy between society and climate, the climate change adaptation framework provides little room for climate and social institutions to coproduce lived environments and there is a strong tendency within this narrative to marginalize questions of power and (agricultural) production within climate change. Doing so leads to a simplistic understanding of 'climate change as a series of external shocks and disturbances to an otherwise coherent society' (Taylor, 2015). Instead climate change can be seen as a series of tensions 'in the way that meteorological forces are actively worked into the production of our lived environments' (Taylor, 2015). Political ecology argues that rural people's contemporary experience climate change needs to be understood within the broader historic context of a rural transformation. 'Without linking localized expressions of ability to broader, historically formed structures of power and privilege, the idea of adaptation can act as a fundamentally de-politicizing concept that produces complex and contested social ecological relations to an abstract appeal to defend communities from external environmental disturbances and threats' (Taylor, 2015).

Political ecology is concerned with how different forms of power are produced and operate across spatial scales that facilitate some actors to influence, profit from and find security, while others are disempowered, marginalized and made vulnerable within the context of ongoing social environmental transformation (Blaikie *et al.*, 1994). Political ecology views adaptive capacity and vulnerability as 'expressions of complex social ecological relations between social class and gender in which such social agents' activity equally seek to transform the lived environment in a historical context' (Taylor, 2015). Taylor (2015) introduces the concept of 'lived environments' to emphasize the social relationships that form landscapes. Climate is therefore in part socially co-produced and the same climate trends can manifest themselves in radically different ways in two different locations and for different land users and social groups.

Political ecology views climate change as intrinsically political

The analytical framework of climate change adaptation is inherently political because it 'shapes how claims can be made, who can speak with authority, who can produce knowledge on the half of whom, who is fronted and entrusted with solving the issue that is identified and who should live with a consequence of the issues it marginalizes' (Taylor, 2015, p.51). In spite of the obvious political nature of climate change adaptation, however, the majority of academic and development agency literature on climate change does not address this issue. In fact climate change adaptation is commonly presented as a technical issue and standing outside of politics. Climate change adaptation is therefore intrinsically a political process because adaptation is closely linked to questions of power and production.

Institutional bricolage

Institutional bricolage, in contrast, goes a step further to address the relationship between communities and their internal environment, a process in which actors consciously or unconsciously reshape or piece together different institutional arrangements at hand (Cleaver, 2001; de Koning, 2011; Cleaver and de Koning, 2015). It consists of adaptive processes in which people ascribe meaning and authority to configure rules, traditions, norms and relationships by altering legislations, ownership or law enforcement through reinterpretation, recombining different institutional elements, negating some institutions and rearticulating others from socially-embedded beliefs. In the process of institutional bricolage institutional components are re-used, reworked and refashioned to perform new functions (Galvan, 1997; Cleaver, 2001; de Koning, 2011; Cleaver and de Koning, 2015).

Rather than analyse institutions in terms of success and failure, institutional bricolage looks at what happens in practice at the local level: how introduced institutions are reshaped in practice by actors who deal with everyday challenges and respond to changes in the social field. It offers new insights into the working of institutions, the role of agency and the constitution of practice often overlooked by mainstream institutional approaches, and simply categorized in terms of success and failure, without conceptually explaining underlying processes (de Koning, 2011; Cleaver, 2012). Institutional bricolage negates the idea that designed institutions can be universally effective in achieving a predefined purpose. As such, it considers institutions to be multipurpose and embedded in practice.

The analytical focus of bricolage is the day-to-day actions of institutions rather than their formal rules and regulations. Bricolage analyses how new institutions are reshaped in practice by actors who deal with everyday challenges and respond to changes in the social field. The bricolage analysis contributes to mainstream institutional analysis by emphasizing the constitutions of practice and associated agency, which is often discussed in terms of success and failure, without conceptually explaining underlying processes (Shering, 2009; de Koning, 2011; Cleaver, 2012). Bricolage analysis considers institutions to be multipurpose and embedded in practice. The analytical framework analysis institutions are unique and thereby challenge the understanding of externally introduced institutions as universally effective in achieving a predefined purpose.

Findings

Climate change adaptation as a political platform for election to district council

Climate change became an important political platform during the 2007 electoral process in Amuria District. Election of councillors for local government took place during the height of the 2007 flood in Amuria District. The major contest (Chairman of the District Council) was fought between the incumbent chairman from the opposition and one from the ruling NRM party. The Chairman had moved from Katakwi District (the mother district from which Amuria was created) to Amuria 'out of frustration over the low service level and accountability of local government towards rural people'.[1] In 2007, the Chairman and the opposition party managed to campaign effectively on platforms that criticized the service level of local government in general and in particular the lack of response from central and local government to food security and upheavals caused by climate stress.

The Chairman won the election as district chairperson with the mandate as the elected leader of people in the district. One of his first actions as Chairman was to establish a verification commission to investigate the causes of poverty in Amuria District. The commission went round to all sub-counties of Amuria districts and collected people's statements on the root causes of vulnerability in Amuria. The commission produced a report that documented in detail the causes of poverty and low level of services in Amuria district. The report underpinned the linkage between the recent wars (fought between government and rebels, as well as involving cattle rustling) and vulnerability of the communities. The report also highlighted poor management of natural resources as its major focus and showed how grazing of local cattle in the wetlands was undermined by government support for expansion of rice cultivation. Owing to the perceived sensitive nature of the report, the District Security Officer under the Resident District Commissioner (the representative of central government) prohibited the launch of the report and ensured that the report was never published. The Chairman of the District Council reacted by bringing newspapers and television producers from Kampala to Amuria and held a dramatic press meeting in which he used the findings in the report to illustrate the effects of floods caused by climate change and a publicly criticized government's inaction, in particular the Minister for Disaster (who hails from Amuria District), for failure to address the plight of the victims of floods.

One effect of the commission and the political turbulence created was that the Chairman and the District Council became more focused on natural resource management, the environment and the role that wetlands play as a buffer against the floods and therefore its direct links with vulnerability in Amuria. As a result the district council requested that the environmental officer develop an environmental ordinance.[2] This ordinance was passed by the district council and later confirmed by the attorney general making it a law. The environmental ordinance called for the protection of wetlands and the development of wetland management plans. In addition, the environmental ordinance prohibited the cutting of endangered tree species (shea nuts, tamarind species) widely used for charcoal burning.

One outcome of this environmental ordinance has been the formulation of an implementation plan for the protection of wetlands (Interview with a Amuria District Environmental Officer, 2013). This was developed through a participatory planning process involving all local users of the wetlands, including those keeping livestock, cutting grass for thatching, collecting firewood, brick making, as well as those cultivating rice. With a view to involve all stakeholders, a team from Amuria District Council, headed by their district's environmental officer, visited all parishes (administrative unit between sub-county and village) located adjacent to Omunyal Wetland in 2010–2011 (the process of implementing the wetlands management plan is described in detail in the next section).

The chairman of the district council often participated in the community planning meetings, in part to show to people that the planning process was supported by the district council. This was particularly important when the demarcation process was opposed by local influential rice growers. In part, the chairman used the meeting to gain political support.

Climate change, once again, came on the political agenda on Amuria District Council in 2010, as the chairman of the council (who belonged to the opposition party) used the inadequate action of central government following the recurrent floods and droughts since 2007 as a political platform. This created debate, partly because the Minister for Disaster at the time was a native from Amuria District and therefore perceived that the criticism from the District Chairman was undermining this role.

The political process of legislation has shaped the mechanisms for the implementation of the laws governing use and protection of swamps that are the main sinks for the control of floods in Amuria. Policy dialogue involving communities created political capital for the local leaders involved. Stiff political competition with national level leaders has, however, been inevitable when realizing that this is an important political niche they may have overlooked.

The electoral process has ensured that the issue of climate change becomes a major policy stream at this particular time largely owing to the effects it has on a wide majority of people/voters. This attracts a high level of policy entrepreneurship among politicians and allied NGOs and agencies.

By the time of the next general elections, in 2011, the district leadership had become very popular through their innovative ways of handling the wetland demarcation process. In 2010, the Chairman of the District was voted by an independent civil society body as the second best District Chairman in the whole country. This was in recognition of, among other things, foresight in community mobilization in the face of climate stress and civil strife. Unfortunately, however, this created a bitter rivalry between the District Chairman and the national politicians from the ruling party, which ended with the Chairman losing his seat. The implementation of the ordinance is still on course, however, because most of the local councillors who subscribe to this process managed to return as representatives to the District Council.

Understanding the 2007 floods in Teso region: external extreme climate change event or co-produced 'lived environment'

The political attention given to climate change by Amuria District Council provided a radically different view on the interaction between climate and society from the mainstream IPCC narrative about climate change adaptation discussed in the analytical framework. Instead of viewing climate as something that is external to society, the causes and effects of the 2007 flood were analysed and discussed during the WMP among all stakeholders who were active natural resource users within the living environment of the Amuria wetlands (Taylor, 2015).

These consultations led to a deeper understanding of the likely causes and effects of flooding in Teso Region. Firstly, the increased intensity and volume of water flowing into Teso Region is understood as an effect of the combination of, on the one hand, extreme intensity of rainfall events in the hills of Karamoja Region caused by climate change and, on the other hand, degradation of vegetation (grassland and tree cover on slopes) in the upper catchment causing enhanced runoff. Secondly, the consultation led to an acknowledgement of the role played by the wetlands and manmade dams in serving as buffers/sinks for the water once it reaches Teso Region. A survey by the environmental department of Soroti District reveals that the natural buffers comprise close to 200 wetland and swamps located across Teso Region. During the participatory planning of WMP, many came to realize that the traditional low intensity uses, including livestock grazing, fishing, cutting straw for building, burning charcoal for home consumption and so on, are conserving the wetlands as a buffer for floods, whereas the intensive cultivation of rice and other commercial crops are reducing the wetlands capacity as a buffer.

The consultations moreover made older people recall how colonial and post-colonial Ugandan authorities managed annual inflow of water from the mountains to the swamps of Teso sub-region by constructing and maintaining

170 dams and forcefully limiting human settlement in low-lying communal pastures. Meso-level institutions are today failing to deal with inflow of water, a problem aggravated by very heavy rainfall and little vegetation to slow the flow of water. The dams are no longer providing a buffer because they are filled with silt and settlement is increasing in former communal pastures as an effect of pressure on land resources. Although the consultative process created a focus on the need to de-silt the dams located in Amuria, the options for the district council to solve the issue are very limited because it is the responsibility of central government, the only institution with the necessary funding, expertise and equipment.

Formulation and implementation of Amuria Wetland Management Plans

The main practical outcome of the environmental ordinance until today has been the formulation of and implementation of the Omunyal Wetland Management Plan (others are the Akare and Alakaituk Wetland Management Plans). These plans were initiated, formulated and implemented by the district environmental officer and his team under the political auspices of the district council chairman. The current district environmental officer started 6 months before the 2007 flood and started working under the new enhanced political mandate for natural resource management discussed in the previous section (Interview with a Amuria District Environmental Officer, 2013).

One of the first tasks of the new district environmental officer was to assess the performance of the existing sub-county environmental committees and parish environmental committees that were established as the local implementing institutional framework linked with the national environmental policy of 2002 (NEMA, 2002). The district review team found these institutions to be poorly functioning and not enforcing protection of the wetlands (Interview with a Amuria District Environmental Officer, 2013). Three groups of farmers encroaching on wetlands can be identified: (i) well-off farmers cultivating rice; (ii) poor farmers who are cultivating traditional crops in wetlands because they have nowhere else to cultivate (a section of the rural population who have insufficient access to land to support their family); and (iii) farmers who cultivate on wetlands as a climate change adaptation strategy because gradual climate change is making agricultural seasons less predictable.

The survey conducted by the district environmental officer revealed that the key reason for wetland encroaching in Amuria was well-off influential farmers cultivating commercial rice. The members of the sub-county environmental committees were unable or unwilling to take on a conflict with this group of farmers. Parish environmental committees and sub-county environmental committees were shown to be compromised by local elite farmers and not effective in protecting the wetlands.

With the view of understanding how to improve local institutions for environmental protection, the district environmental officer and his team carried out a stakeholder analysis that mapped the many uses of wetlands. These are young men herding livestock, women cutting grass for thatching, women fishing, women collecting firewood, men making bricks for sale and men cultivating rice for sale. The exchange with local community also revealed that the majority of the rural population was in support of wetlands conservation and that the encroachers are a small but influential minority. On the basis of these studies the district environmental officer and the district council established an argument for why demarcation of wetlands could conserve them as sinks during extreme floods. The 2007 floods made district authorities realize that wetlands serve an essential role as sinks for the floods and establish a longer-term climate change adaptation argument for the need to demarcate wetlands.

After the 2007 floods, the District Environmental Officer received a clear political mandate from the District Council to develop environmental ordinance, as the legal basis for developing a wetlands management plan.[3] The political resolve to act on the environmental issue was reinforced by an incident in 2009, where the natural resource use conflicts within the wetlands had escalated into violent struggle where the police had to intervene to end it. Environmental ordinance was ready from the technical staff in 2009. During the year-long process of amendment and acceptance by District Council and Attorney General, the District Environmental

Officer and his team 'sensitized' the community through a series of meetings aimed to raise awareness about the importance of wetland conservation and to collect people's views and perceptions of how this could be done.

By the time the environmental ordinance was finally adopted by the District Council and made into law in 2010, the District Environmental Officer and his team were ready to develop a wetland management plan for Omunyal Wetland, which is the most important of four waterways into Amuria District from Katakwi Region. Thereafter other wetland management plans were also developed which resulted in the demarcation of Akare wetland in Abarillela sub-county and Alakaituk wetland in Asamuk sub-county. These wetlands' management plans were developed through a participatory planning process involving all local users of the wetlands, including livestock herdsmen, women specializing in cutting grass for thatching; women in general collecting firewood; men making bricks for sale; men and women fishing; as well as men cultivating rice. These groups of users were not organized, however, the participatory planning process made the different users of the wetland clearer to the community.

With a view to involve all stakeholders, a team from Amuria district council, headed by their district's environmental officer, visited all Parishes (administrative unit between sub-county and village) located adjacent to Omunyal Wetland in 2010–2011. The purpose of the repeated visits was to involve all different users of the wetlands in a participatory process that made clear to the community who was using it for what.

These meetings revealed that the central conflict of land use in Omunyal and other wetlands was between the traditional uses of livestock keeping, cutting grass for thatching, collecting firewood, brick-making and fishing on the one hand, and commercial rice cultivation on the other. Expansion of rice cultivation into the wetlands had been made possible by a combination of several factors, the most important being that rice cultivation is profitable and carried out by well-off and influential farmers. It had been possible for this group of farmers to encroach into the lands that had previously been used for a range of traditional uses because of unclear demarcation between the wetlands and drylands. An additional reason for the expansion of rice cultivation is JICA, a national rice support programme that promotes indiscriminate expansion of rice cultivation. The programme provides high-quality seed varieties of both lowland and upland rice varieties. Participatory planning process reveals that those using the wetlands in the traditional and more sustainable manner far outnumbered the farmers expanding rice cultivation into the wetlands. It was therefore relatively easy for the team to make the majority of the stakeholders agree on a plan to protect the wetlands.

At the moment, a fully participatory and accountable system for wetlands protection has been instituted. The system is regulated by whistle-blowers identified and living among the communities in Amuria. The acceptability of the system is driving other communities whose wetlands have not yet been demarcated to voluntarily demand a fast demarcation.

Institutional dynamics of land demarcation

The wetland management plan was institutionalized in 2012 and provides a detailed border setting, with clear demarcation of where the wetlands end and where the dryland starts. To enforce the wetland management plan, a fully participatory and downwards accountable system for wetlands protection has been instituted. At the grassroots level, the participatory planning process establish a network of volunteer whistle-blowers who live in the community, linked to the sub-county Environmental Office/ sub-county environmental committees. These whistle-blowers act as monitors and report violation of the wetlands use to the sub-county environmental committee. The previously relatively dormant sub-county environmental committees have been revitalized and given power to report encroachers to the police. As part of the implementation of the Environmental Ordinance, one police officer from each police post received environmental training and were selected to be focal persons for the wetland management plan in the sub-county. The district has budgeted for bi-annual district monitoring teams, which are composed of members of the sub-county environment committees and technical staff, where the violators are verified and fined.

The participatory character of the wetlands management plans has enhanced the respect for the district environmental officer. He is using this enhanced legitimacy to (together with sub country environment committees) enforce associate conservation measures that he previously was powerless to address, such as setting up buffer zones to avoid creation of channels and water abstractions in the wetlands, prohibit seine baskets that trap all categories of fish and avoid creation embankments in the wetlands. This allows other traditional activities carried out in the wetlands like fishing, cattle grazing, hunting and harvesting of materials for crafts to continue. In spite of this new legitimate and widely agreed upon demarcation line and wetlands management institutions, the communities that share the wetland resources continue to have different values, providing the potential for resource conflicts. The implementation of the wetland management plan has shifted the relative power relations between different social groups. Rice farmers continue to oppose a general application of the 30 m buffer zone for crop production. Meanwhile, livestock keepers, a group that has clearly gained from the demarcation, are allowed to penetrate deep into the centre of the wetlands for grazing and watering their animals because this is regarded as sustainable land use practice.

Community representatives in the sub-county environmental committee have limited capacity and will to negotiate with local government and community land-use groups, when encroachments are made by bigger and more influential society members. Observations and qualitative interviews with key informants reveal that whistle-blowers are selective and only reported the less influential members of the community, whereas encroachment by members of the local elite was overlooked to avoid potentially damaging conflicts. The influential members of the community are also treated lightly by the police. When the sub-county environment committee report influential members to the police for encroaching, they are often frustrated by the magistrates that prefer to use weaker colonial laws to prosecute encroachers (instead of the new Environmental Ordinance). In most of such cases, fines given to encroachers are too low to act as deterrents. Some well-connected encroachers are never prosecuted and others are released on bond pending investigations that never end (interview with key informants). A recent trend is that the sub-county management committee, in some cases, avoids reporting to the police but handles encroachments at the local neighbourhood level with negotiated punishments between encroachers and enforcement officers.

Discussion

The experience in Amuria District illustrates that it is possible to put climate change on the political agenda of local government. The study shows that the process was driven by a charismatic individual in opposition to political establishment who used the widespread concerns over climate change among the electorate as part of his political platform. Although other local government politicians view climate change as an issue for which it is difficult to gain votes, this politician campaigned by successfully articulating average farmers' everyday struggles with issues of poverty and natural resource management. After the election victory, the new District Council Chairman insisted on acting on environmental issues in spite of the absence of budget, ordering the environmental officer to develop a participatory plan of action.

This experience indicates that political initiatives at the District Council level, including enacting an environmental ordinance and formulating and implementing wetlands management plans, can create an enabling environment for climate change adaptation. The successful formulation and enactment of an environmental ordinance proved to be a viable policy instrument for local government. The involvement of all stakeholders in the formulation of wetlands management plans, under the auspices of the environmental ordinance, ensured that the plans had legitimacy in the eyes of the majority of the rural people. Popular participation in formulation of the wetland plans, involving all users of the wetlands, made it transparent for all that the rice growers were a minority group. By participating actively himself, the political champion gave political clout to the process, making it possible to disempower the local elites involved with rice cultivation and make plans that protect further encroachment in the wetlands and ensure sustainable plan use practices. During about half of the participatory planning sessions, the district used its authority to manipulate the outcome to the dissatisfaction rice cultivators. The participatory planning approach ensured that the demarcation line between cultivated land and wetlands was based on a broad agreement among the majority within the local society and thereby likely to withstand challenges from local elite interested in commercial rice cultivation.

The implementation of the wetlands management plans resulted in new content of existing institutions and the creation of new institutions through a process of institutional bricolage. The Parish and Sub-county Environmental Committees were revitalized and took on the challenge of monitoring the extent to which wetland users adhere to the agreed demarcations and regulations. Parish and sub-county Environment Committees have gained new roles and responsibilities. These committees that were historically subject to elite capture have now changed roles and become more accountable to the wider community. By creating greater awareness and linking committees with volunteer whistle-blowers, and by using their reports in bi-annual monitoring exercises, bureaucratic processes have changed, with committees being more accountable and more effective in controlling the wetlands.These committees also interact with new actors, namely the whistle-blowers and the police. The police have multiple roles as both regulation and enforcement and are also trained to participate in the wetlands demarcation and in facilitating the work of whistle-blowers in the community.

The aggregation of government roles, local norms, and local beliefs that can be seen to be giving voice to new actors in wetland management, i.e. giving greater space to women and poor farmers, who under the old regime were threatened by commercial rice production. The study also shows, however, that enhanced political invited space and accountable institutions cannot alone empower people. The absence of financial resources and less political interest and support from local government after the District Chairman was replaced at the 2011 election has reduced the effectiveness of the new institutions in enforcing the restrictions on the encroachment of wetlands.

Notes

1 Interview with Mr Julius Ocen, Chairman of Amuria District Council 2007–2011.

2 Amuria districts 2010. Environmental Ordinance.

3 Interview with former district council chairman of Amuria District.

References

ActAlliance (2010) Assistance to people affected by floods in Teso sub-region, North Eastern Uganda. *Alert. Uganda* – No. 32/2010. Available at: http://reliefweb.int/sites/reliefweb.int/files/resources/86C160841D32064B852577B90054B66A-Full_Report.pdf (accessed 10 August 2016).

Adger, W.N. (2006) Resilience, vulnerability, and adaptation: A cross-cutting theme of the International Human Dimensions Programme on Global Environmental Change. *Global Environmental Change* 16, 268–281.

Bashaasha, B., Mangheni, M.N. and Nkonya, E. (2011) Decentralization and rural service delivery in Uganda. *IFPRI Discussion Paper 01063*. International Food Policy Research Institute, Washington DC.

Bassett, T.J. and Fogelman, C. (2013) Déjà vu or something new? The adaptation concept in the climate change literature. *Geoforum* 48, 42–53.

Blaikie, P. and Brookfield, H. (1987) *Land Degradation and Society*. Methuen, London.

Blaikie, P., Wisner, B., Cannon, T. and Davis, I. (1994) *At Risk: Natural Hazards, People's Vulnerability and Disasters*. Routledge, UK.

Cleaver, F. (2001) Institutional bricolage, conflict and cooperation in Usangu, Tanzania. *IDS Bulletin* 32, 26–35.

Cleaver, F. (2012) *Development through Bricolage. Rethinking Institutions for Natural Resource Management*. Earthscan from Routledge, UK.

Cleaver, F. and de Koning, J. (2015) Furthering critical institutionalism. In: Cleaver, F. and de Koning, J. (eds) Challenges of critical institutionalism. *International Journal of the Commons* 9, 1–18.

de Koning, J.D. (2011) Reshaping institutions – Bricolage processes in smallholder forestry in the Amazon. PhD thesis. Wageningen University, The Netherlands.

Epilo, J. (2011) Climate change challenges to crop production, adaptation and mitigation strategies for livelihood sustainability in Teso Sub-Region, Eastern Uganda. MSc thesis, Faculty of Life Sciences, University of Copenhagen, Copenhagen.

Friis-Hansen, E., Bashaasha, B. and Aben, C. (2013) Decentralization and implementation of climate change policy in Uganda. *DIIS Working Paper 2013:17*. Danish Institute for International Studies, Copenhagen.

Galvan, D. (1997) The market meets sacred fire: land pawning as institutional syncretism in inter-war Senegal. *African Economic History* 25, 9–41.

IPCC (2007) *Climate Change 2007: Impacts, Adaptation and Vulnerability, Contribution of Working Group II to the Fourth Assessment Report 2007*. Cambridge University Press, Cambridge, UK.

NEMA (National Environment Management Authority) (2002) *National State of the Environment Report for Uganda 2002*. National Environment Management Authority, Ministry of Natural Resources, Government of Uganda, Kampala, Uganda.

Robbins, P. (2004) *Political Ecology: A Critical Introduction*. Critical Introductions to Geography. Blackwell, Oxford, UK.

Shongwe, M., Jan Van Oldenborgh, G. and Van den Hurk, B. (2010) Projected changes in mean and extreme precipitation in Africa under global warming. Part II: East Africa. *Journal of Climate* 24, 3718–3733. American Meteorological Society. doi: 10.1175/2010JCLI2883.1.

Taylor, M. (2015) *The Political Ecology of Climate Change Adaptation*. Routledge, New York.

World Bank (2006) *Development and Climate Change*. World Bank, Washington DC.

ge courtesy of Tine Enged

8 Political Ecology of Climate Change Management in Rural Uganda

Charles Aben

The Meaning of the Political Ecology Concept in the Context of Rural Uganda

As in other parts of the world, the recent wave of climate change has become a challenge to communities across Africa, besides putting valuable resources such as wetlands at risk of degradation. The political ecology concept seeks to explain how climate change has altered the traditional access rights of community members, thereby increasing the vulnerability of those dependent on them for grazing, fishing and other low-intensity traditional activities. The concept also analyses the extent to which changes in resource protection mechanisms involving a series of compromises and relaxations of rules affect the sustainability of the Awoja watershed. The concept also seeks to determine how changes in the watershed have altered decision-making processes and governance linkages between traditional leaders, the decentralized national leadership and other stakeholders such as the non-governmental organizations (NGOs) involved in the management of climate change. Changes in the benefits and losses to the local community, leaders and neighbouring communities within the watershed may also create significant variations in the level of vulnerability amongst the different community members and act as a basis for emerging socio-economic and socio-cultural interests and resource conflicts.

In this respect, the contribution of political ecology is to help us think critically about how the framework of climate change management produces the issues that it subsequently seeks to resolve. This situation was identified by Taylor (2015) when he revealed the existence of important transformations during processes of adaptation to climatic change across varied spatial scales, with challenging implications for humanity.

With regard to institutional and community interactions, the political ecology of climate change within the watershed concerns changes in processes of access, rights of ownership and the local government's response to the ecological effects of climate change. This also concerns a critical examination of how natural resource governance systems shift in circumstances when government agencies, several NGOs and local communities are engaged in various activities, including the demarcation of the wetlands, especially in Amuria, and the demarcation of the uplands, particularly for wildlife conservation in Magoro sub-county in Katakwi District, mechanisms the government believes will guarantee the sustainable utilization of resources and innovations that are responsive to the effects of climate change.

The chapter also draws its understanding of the term 'political ecology' from analyses of changing political processes, social issues, economic issues and how they interact to influence stakeholders' engagements, flexibility in implementing natural resource regulations, power relations, opportunity structures, resource access and use, and the management of mobility as an adaptation to climate change. Understanding the concept of political ecology therefore opens up a deeper set of political questions about power, the co-production of hazards and solutions and environmental change than is possible within the paradigm of climate change management.

It should be noted that since the 2007 floods in the Teso region, combined with the drought in the Karamoja region, virtually no part of the Awoja watershed has remained untouched by the community's compulsion to transform nature and livelihoods in order to survive in changing climatic conditions. Interventions by local government institutions, the donor community and changing local commercial interests have become the source of intense and bitter social struggles across the watershed.

The management of climate change in the watershed and its effects have generally created a new order characterized by changing political, social and economic issues. This situation is similar to that encountered by Moore (2010) and Barbier (2011) when they concluded that climate change appears to present a tremendous challenge to modern practices in which nature manifests itself no longer as a passive resource but as a dynamic historical agent with the potential to dramatically shape humanity's future. Therefore analysis of the interactions within the watershed indicates that climate change creates a new order of relationships that are characterized by profound vulnerability and anxieties. Thus climate change management catalyses a determined human response to the threats posed by climate change. This represents a different order that has been referred to as the political ecology of climate change management (Adger *et al.*, 2009). Climate change management has drawn enormous efforts from national and international governments, putting international agencies, governments, corporations, NGOs and social movements into a common and encompassing framework (Ireland, 2012). In the Ugandan context, the political ecology of watershed management is contingent on decentralization processes at the local government level, where the management of natural resources in a changing climate is embedded in the political, economic and social processes at the district and sub-county levels. Understanding the relationships between socio-cultural and socio-political forces in the watershed is important in unpicking how local communities are impacted by the emerging rules and regulations and how they are responding to the effects of climate change.

Cultural and Political Issues in Relation to Natural Resource Management in a Climate Change

In the context of the Awoja watershed, an analysis of the interface between culture and politics is very important. This is mainly because the cultural structures could emerge from the climate challenges and become stronger. This has the potential to shift the power to allocate and control the use of resources sharply, as vulnerable populations are bound to lean towards their traditional culture rather than the political structures of the local governments.

A good example in the watershed was when, instead of increasing the deteriorating soil fertility, the population of Amuria, Katakwi and Soroti adjusted their farming systems to fit the deteriorating soils. It could also explain how the people of Angisa established parallel committees to solve community problems alongside the existing local government structures. Whenever there was a criminal act in the community, instead of reporting the matter to the police, it was reported to the clan chairman (conversations with leaders of Angisa in Katakwi). This represents a great politico-cultural shift that has the potential to render the existing political structures irrelevant in the eyes of the people. This relationship has cascaded down to other socio-political elements that have made governance difficult because of a lack of trust in the prevailing governance systems because people trust their clan systems more. This could be a livelihood strategy because the clan system becomes the controller of the allocation of and access to the resources in the area rather than the government. Community members had found that they needed to work closely with their clan leaders for purposes of livelihood protection. This view of the response to climate change diverges slightly from those found by Blaikie (1985), Blaikie and

Brookfield (1987) and Robbins (2012), in which they noted that, for the early political ecologists, rather than seeing changes in livelihoods as cultural, changes are seen as a response to marginalization by capitalism and social exclusion that puts some groups of people at a disadvantage in terms of resource access. It should be noted that the strength of culture against the political powers implies that the influence of culture in the process of adaptation to climate change has been strongly entrenched and has sustained irrational forms of land management resulting from traditional values that have complicated the process of adoption. It is important, however, to note that a failure to weaken traditional practices and their associated belief systems might leave the watershed societies trapped in a state in which resource use remains inefficient and prone to degradation and depletion in the face of population growth.

While this situation exists, it should be clearly understood that value systems offer special resistance to change and that this is facilitated if the material economic environment in which they can flourish is destroyed or weakened. Within the Awoja watershed, the value systems, especially the traditional values, are so strong that people respect their traditional leadership more than the leadership of local government. The study finds that even local government leaders still have strong inclinations towards traditional forms of leadership and were not struggling to alter beliefs. This has made it difficult for them to implement a number of regulations, especially those regarding demarcations, because they were highly sympathetic to the population and feared being reprimanded by the clan leaders if they acted centrally.

If the views of Piera Blaikie (1985) are considered in the context of this chapter, it can be deduced that the inconsistencies of political ecologists in relation to cultural ecology's adaptation analysis stem from its marginalization of the historical dynamics that have shaped agrarian environments in Awoja watershed. Instead, political ecologists have sought to situate localized processes within causal forces. It was for this same reason that the early political ecologists believed that, by sticking to culture, rural communities always betrayed the progressive land management changes of the postcolonial period, thus providing capitalism with an opportunity to take over. The same challenge was observed in the Awoja watershed in which big commercial rice farmers were taking over big chunks of land from the rural farmers through a process of connivance with and bribing of the clan leaders.

If the situation in the Awoja watershed is viewed in light of the position of Peet and Watts (2004), it can be seen that the emergence of big commercial rice farming and capitalism in Awoja is by no means a new occurrence. As Peet and Watts (2004) put it, 'market integration, commercialization and the dislocation of customary forms of resource management in place of adaptation became the lodestars of a critical alternative to the older cultural or human ecology'.

It should be noted that an analysis of the manifestations of political and cultural ecology within the Awoja watershed means that any effective programme to manage climate change must take both aspects into account because they co-exist within the context of climate change management. Therefore, in analysing natural resource management in rural Uganda, one cannot avoid using cultural ecology alongside political ecology perspectives in understanding the influence of cultural value systems, which can lead to changing livelihood systems.

Context of Climate Change Governance within Awoja Watershed in Eastern Uganda

The Awoja watershed system is best remembered for the widespread flooding of 2007 that was greater than experienced before. The cause of the floods was basically the rain flowing from the hills of the Karamoja region. This region consists of the districts of Nakapiripirit, Moroto, Kotido, Napak, Abim, Kabong and Amudat, all highland areas. This has subjected the watershed to floods whenever it rains as rainwater flows from the Karamoja region to the Awoja region.

In 2007 the floods affected the road infrastructure, food security establishments and livelihoods, and caused the mass displacement of people in the Teso sub-region. This created a situation with climate refugees in these neighbourhoods, which has proved to be the basis for a series of political changes in the management of the watershed, given that a number of communities that did not previously work together have been able to implement some joint agreements for the purposes of survival. Flooding has often forced the Karamajong to move their cattle to the Teso region, on the basis of a claim that they are merely following 'their' water when it runs to Teso in the form of floods. This has always been the major cause of conflict between the two tribes and a reason for civil strife. The situation complicated the adaptation process because communities lived in fear and could not stay in some places, especially in the Angisa region in Katakwi District, where people were displaced and had to seek refuge in displaced persons' camps. The watershed has experienced an unprecedented degradation of resources, especially of the wetlands that are being used as livelihood alternatives owing to the extended drought periods that have left people with no food because the uplands are no longer productive. The interesting aspect of the Awoja watershed case is that, while it is an area affected by floods, it is at the same time an area greatly impacted by drought because the floodwaters originate in a different far-off region. In recent years, the whole state has been affected by fluctuating water levels arising from droughts and increasingly torrential rains that have complicated natural resource management in the area. The responses of local institutions to the floods and droughts seems to vary across local government systems because each community has been affected differently from every other owing to the different demographic characteristics and geographical terrain. The state of the Awoja watershed is generally an example of how humanity has moved to the pursuit of survival, and it represents a rare interface between the pursuit of rationality and efficiency. Hence, this has created a state of unparalleled access rights to resources and consumption, as well as leaving a trail of resource depletion, land degradation, environmental pollution and the extinction of certain resource species, especially rare birds. In the Awoja watershed, this holds true because there has been a struggle between the traditional leaders and the land-use practices desired by communities, along with conflict between the state and the communities in the implementation of the land-use policies. The parallel states are evident where the government has established policies, especially regarding land demarcation, and the communities have continued to encroach on the demarcated land in the interests of survival.

Within the Awoja watershed, climate change seems to pose a different order of questions for governance, especially concerning land-use regulations. It should be noted that the use and abuse of the watershed continues unabated owing to a lack of alternative livelihood strategies, possibilities and opportunities, itself due to the increasing population growth in the watershed and selective implementation policies protecting the environment pursued by the local leaders. This is closely connected to their sympathy for the local communities, the lack of essential services, especially extension services, and other constraints that often appear to be relatively localized and constitute permeable limits to human designs. In the context of the Awoja watershed, however, the idea of climate change seems to represent a much greater challenge, one so embedded in cultural contexts that it is not easily observable. This is the challenge of the increased power of the traditional leaders in land-use management, whom the community members respect more than the government and its policies. This has continued to jeopardize the effectiveness of climate change management efforts, and the negative effects of climate change are thus likely to increase if this problem is not handled properly.

During the study of the Awoja watershed, the governance of climate change management was found to be the responsibility of many stakeholders. There is a decentralized framework for environmental management in which the local government environmental committees are supposed to monitor and implement the environmental management regulations. This begins in the local councils, where there is a secretary for the environment, whereas the parish and sub-county levels also have their environmental secretaries. These are supposed to work hand in hand with the district environmental officer to implement environmental policies within the districts. In the Awoja watershed, however, the community has created its own parallel committee that is actively involved in monitoring resource use in the area. The committee even has its own parallel regulations on land use that are different from those in the national framework. These include prescribed punishments for failures to obey the land use policies that are laid down. The other stakeholders in climate change governance were the traditional leaders and their clan management committees, which worked in consultation with NGOs, the local councils and the community environmental committees. It should be noted that, as a result of this governance framework, the challenges of climate change in the area have turned out to be more a matter of the co-production of hazards, with each party struggling for possible benefits, something that has created a situation of winners and losers in climate change management. It is therefore important to note that the politico-ecological aspects of the governance of climate change changed a great deal owing to the greater involvement of NGOs in the social protection of the affected communities and in ecological reconstruction for purposes of sustainable climate change management. The focus of government has been political. For instance, demarcation of the wetlands has proved to be more about the need to reduce community conflicts in the use of the wetlands than about conservation needs. This explains why, in some districts like Amuria, demarcation of the wetlands has succeeded, whereas in Katakwi and Soroti it has failed because in these districts conflicts over land use have been minimized, a consequence of the cattle herders and cultivators having agreed to co-exist. The community members, in contrast, are more focused on survival because climate change has had significant impacts on the condition of the land as a resource, and in most cases the yields in upland areas have declined considerably. People have now turned to protected areas such as the wetlands for survival and have tried to demand policies that are purely survival related.

Vulnerability and Marginalization in the Awoja Watershed

From the Awoja context, which is purely an agrarian environmental setting, adaptation to climate change can be viewed as representing a case of communities adjusting to climate threats. However, attention should also be paid to the historical roots of the vulnerability that many marginal user groups within the watershed face. In this respect, core questions concerning the history of and control over communal land, water resources and forests within the Awoja watershed and the traditional mechanisms for the distribution of risks and rewards within the system need to be addressed. This is because power relations may determine how different social groups experience climate change and structure their relative abilities to respond accordingly. It is perhaps with this in mind that political ecologists such as Peet, Robbins and Watts (2011) have criticized mainstream approaches as a 'shopping list of "conditions" for adaptive governance' rather than an analysis of the complex political, cultural and social dynamics at work. O'Brien *et al.* (2010) and Pelling (2011) have focused instead on pre-existing forms of social differentiation as a vector of vulnerability to climate change. An emerging body of literature examines vulnerability from a broadly 'human security' perspective, emphasizing how those worst affected by abrupt climate change are disproportionately drawn from segments of society that are already chronically marginalized in daily life. In considering relations of power in the Awoja watershed, it becomes clear that environmental and social change does not affect everyone equally. Less resilient communities appear more vulnerable and can be severely affected by change, thus limiting their opportunities for adaptation. It may be more appropriate to postulate that climate change is shaped by a range of social factors, such as the degree of social inequality and political representation.

It can also be agreed from the context of the Awoja watershed that inequality is very important in political ecology analyses of the stratified distribution of assets or access to public resources by members of the community. This also relates to a given and unequal state of affairs and its societal impact in respect of levels of vulnerability or the adaptive capacity of community members. In contrast, a focus on power seeks to understand the dynamic relationships and processes through which humans and their environments are unequally produced over time. This view is also supported by Blaikie *et al.* (1994), Bohle *et al.* (1994) and Oliver-Smith (2004) when they stress that the political ecology perspective aims at examining how different forms of power are produced and operate across spatial scales that facilitate some actors in influencing, profiting from and finding security, whereas others are disempowered, marginalized and made vulnerable within the context of ongoing socio-environmental transformations. A similar situation is evident in the Awoja watershed, where the main beneficiaries were the cattle herders and commercial rice growers: even though everybody is affected by climate change, some are more affected than others. It was observed that while land in upland areas had become scarce due to overpopulation and the demarcation of wetlands, which caused cultivators to move to the uplands, land was abundant in the wetlands. This has become an opportunity for cattle herders, given that the wetland is not open to cultivators, only to cattle herders, thus making them richer.

From this perspective, the study of vulnerability in a climate change adaptation framework cannot be reduced to identifying the disadvantaged members of society. As revealed in the case of the Awoja watershed, everybody is vulnerable in one way or another because the challenges of climate change are not selective. This difficulty of categorizing vulnerability was also mentioned by Hinkel (2011) when he suggested that categorizing the vulnerable is such a challenge because, in the context of climate change, the groupings are ever changing. In the case of the Awoja watershed, it was found that climate change refugees are forever moving and that this affects their group settings. Owing to the difficulty in the categorization of vulnerability, Mosse (2007, 2010) and Collins (2010) argued that, instead, there is need to focus on the exploration of 'vulnerablization' as a relational process in which vulnerability is produced and reproduced over time between social groups within the active production of their lived environments.

To this end, the field of political ecology examines the power relations involved in both representing and managing livelihoods within the context of the Awoja watershed. The chapter has attempted to explain how existing socio-environmental orders influence the distribution of the gains and risks arising from climate change challenges: who has 'power to adapt', how is such power formed and maintained, and at whose potential expense does it operate? This understanding was reflected on by Mustafa (2005) when he stressed that seeking a political ecology perspective requires us to be acutely sensitive to the power dynamics that construct our lived environments and that actively yet unevenly reshape their social and physical landscapes. In particular, it demands that we consider the thorny relational issue of how the insecurity of some might be intimately connected to the relative security of others. This chapter therefore stresses that climate change management in the face of vulnerability is quite challenging and that any policy formulated to deal with it should be all-inclusive so as to take care of all groups in the community.

The chapter indicates that the challenges of rural life are continually co-produced and reshaped by active social and economic forces. It is within this context of ongoing and dynamic changes in management that we can begin to appreciate the past, present and future impacts of climatic change.

Challenges of Climate Change Management in a Rural Context

Environmental change in Awoja seems to be embedded in the struggle of community members to adapt to the effects and economic opportunities presented by a changing climate. This creates a dilemma for governance systems in choosing between promoting development and community survival versus conservation. The commoditization of rice production in the Awoja watershed has tested the level of political will by local governments when it comes to protecting the environment. While in Amuria, this will is evident, as seen from legislation and the implementation of resource management regulations, commercial and political interests, elite capture, inaction and compromises in managing the watershed on the part of the leaders is common in Soroti. In the watershed, adaptation to climate change has accelerated the pace of environmental reproduction. Different agencies are working at different scales and with vastly different concentrations of power, from the level of household livelihoods through to the decisions of bureaucrats under the decentralization framework. Communities and their agencies are therefore agents of the degradation of the watershed. Besides rice production and quarrying, charcoal burning and bush fires are among the activities that are co-producing hazards in the watershed. On the positive side, however, multi-stakeholder involvement in management of the watershed is generating innovations aimed at finding solutions to the effects of a changing climate. Climate change effects are therefore being actively co-produced by the actions of communities and their institutions.

Conclusion

From the above discussion, it can be concluded that climate change management in a rural setting in Uganda is highly contingent on the social, economic and political environments within which local people live. The political and economic processes and social dynamics of the Teso sub-region in the past three decades provides a unique opportunity to use political ecology as an analytical framework to understand adaptation to climate change in Uganda. As Taylor (2015) has mentioned, the political ecology concept reflects the fact that climate is not something 'out there' but is actively produced as an essential moment in the formation of lived environments, the discursive boundaries in which the adaptation paradigm becomes tenuous.

References

Adger, W.N., Lorenzoni, I. and O'Brien, K.L. (2009) *Adapting to Climate Change: Thresholds, Values, Governance*, 1st edn. Cambridge University Press, Cambridge, UK.
Barbier, E. (2011) *Scarcity and Frontiers*. Cambridge University Press, Cambridge, UK.
Blaikie, P.M. (1985) *The Political Economy of Soil Erosion in Developing Countries*. Longman, London.
Blaikie, P.M and Brookfield, H.C. (1987) *Land Degradation and Society*. Methuen, London.
Blaikie, P.M., Cannon, T., Davies, I. and Wisner, B. (1994) *At Risk: Natural Hazards, People's Vulnerability and Disasters*. Routledge, London.
Bohle, H., Downing, T. and Watts, M. (1994) Climate Change and Social Vulnerability. *Global Environmental Change* 4, 37–48.
Collins, T. (2010) Marganilization, facilitation and and the production of unequal risk: The 2010 Paso del Norte Floods. *Antipode* 42, 258–288.
Hinkel, J. (2011) Indicators of vulnerability and adaptive capacity. Towards a classification of science-policy interface. *Global Environmental Change* 21, 699–714.
Ireland, P. (2012) Climate Change Adaptation. Business as usual aid and development as an emerging discourse for change! *International Journal of Development Issues* 11, 92–110.
Moore, J. (2010) 'Amsterdam is standing on Norway' Part 1: The alchemy of capital, empire and nature in the diaspora of silver. *Journal of Agrarian Change* 10, 33–68.
Mosse, D. (2007) Power and durability of poverty, a critical exploration of the links between culture, marginality and chronic poverty. CPRC Working Paper, Anthropology Department, School of Oriental and African Studies, University of London 107, 1–60.
Mosse, D. (2010) A relational approach to durable poverty, inequality and power. *Journal of Development Studies* 46, 1156–1178.
Mustafa, D. (2005) The production of an urban harzardscape in Pakistan: modernity, vulnerability and the range of choice. *Annals of the Association of American Geographers* 95, 566–586.
O'Brien, K., St Clair, A.L. and Kristoffeersen, B. (eds) (2010) *Climate Change, Ethics and Human Security*. Cambridge University Press, Cambridge, UK.
Oliver-Smith, A. (2004) *Theorising Vulnerability in a Globilized World. Political Ecological Perspective in Mapping Vulnerability, Disasters Development and People*. Earthscan, London.
Peet, R. and Watts, M. (eds) (2004) *Liberation Ecologies: Environment, Development, Social Movements*. Routledge, London.
Peet, R., Robbins, P. and Watts, M. (2011) *Global Political Ecology*. Routledge, Abingdon, UK.
Pelling, M. (2011) *Adaptation to Climate Change*. Routledge Press, London.
Robbins, P. (2012) *Political Ecology. A Critical Introduction*, 2nd edn. Blackwell, Oxford, UK.
Taylor, M. (2015) *The Political Ecology of Climate Change Adaptation: Livelihoods, Agrarian Change and the Conflicts of Development*. Routledge, Abingdon, UK.

Image courtesy of Tine Engedal

9 Local Political Processes and the Management of the Awoja Watershed in Eastern Africa

Charles Aben, Esbern Friis-Hansen, Jacob Agea and J.J. Okiror

Introduction

Stakeholders have recognized the need for effective governance of the Awoja watershed, not only as an environmental protection responsibility, but also as a means of enhancing adaptation to the climate change that has taken place in the area. Climate change has generated the need for government and other stakeholders to come together to provide social and economic protection for the affected communities through adaptation to climate change, as well as to protect important natural resources, including land, water, wetlands and forests, from abuse and further degradation from the very process of adaptation.

Natural resource protection in the Awoja watershed is becoming crucial in safeguarding communities whose livelihoods directly depend upon it. This watershed, located in Eastern Uganda, is an area challenged by frequent droughts and floods. The main aim of this study is to determine the local political processes that influence the management of resources in the watershed in a changing climate. The study analyses how governance decisions are made, the influence of political interests on the management of resources, mechanisms for dealing with degradation, mechanisms for dealing with non-compliance with regulations, and how stakeholder participation influences power relations in the watershed. The Awoja wetland system is best remembered for the widespread floods of 2007, which affected the road infrastructure, food security and livelihoods and caused the mass displacement of the population in the Teso sub-region. As a result, community members, national environmental management departments and the donor community realized that climate change was a reality in the watershed. This realization influenced a number of demographic aspects, traditional practices and changes in governance that generally altered resource use in this watershed. The changing weather patterns and more extreme weather events are, however, only part of the explanation for this situation, as access to resources is subject to political considerations arising from interactions between communities, meso-level organizations and local branches of the state. According to NEMA (2010) and Mwangi and Wardell (2012), actors typically interact regularly in communities involving both horizontal and vertical interactions.

The study therefore focuses on the political processes involved in the management of the Awoja watershed, and in particular seeks to understand the roles of traditional leaders, local government officials and the community members themselves, since all these stakeholders have become important in the management of the watershed. Therefore, the three main hypotheses of the study are:

1. Extreme climate change events in the form of the 2007 and 2012 floods are causing serious hazards for rural citizens in the Awoja wetlands. However, as a result of the path-dependent institutional behaviour of local government political leaders and administration, the sub-counties are receiving very low political support and little funding to carry out their task of adapting to floods.
2. There is enhanced invited political space where sub-county and local government staff are unable to follow normal planning procedures to carry out their mandate owing to inadequate human and financial support. Therefore sub-county and parish staff have invited other local stakeholders to participate in actions connected to climate change adaptation.
3. The final hypothesis is that the invited/claimed political space has resulted in institutional bricolage among local institutions. Non-governmental organizations (NGOs) are taking on the roles of local government, and new forms of collaboration and new institutional links are emerging.

The Contextual Framework of the Awoja Watershed

Awoja Watershed is classified as part of the Lake Bisina wetland system, which owes a great deal of its ecological existence to Lakes Bisina and Opeta and never dries even during the long dry seasons (Nature Uganda, 2005). However, its sustainability is influenced by the socio-economic activities of 14 districts in Teso, Sebei, Karamoja and Bugisu sub-regions, which experience extreme weather events, including droughts and floods in the uplands. The flat terrain, which is traversed by numerous swamps and other ravine wetlands, makes the area prone to flooding from the high-altitude Karamoja region into the Teso lowlands. The flow of water from the high grounds in Karamoja into these lowlands is a source of conflict over the real ownership of the latter. Although the land is geographically in Teso, the Karamojong pastoralists believe that they have a right to graze and water their animals there because the water originates in Karamoja.

In an average year, annual rainfall totals are typically 1100–1200 mm distributed between two seasons running from March to July and from September to November. Late November to late February or early March is traditionally the long dry season, with another, shorter season from mid-June to late July, but these periods have become variable, with frequent spells of drought causing famine (MWE, 2007). The watershed is used for both growing crops and raising animals in both the wetlands and the uplands. Small-scale farming has long predominated in the area, greatly reducing vegetation cover, this being a likely trigger for negative environmental effects such as floods and droughts, reduced soil nutrients and biodiversity loss resulting from habitat conversion (Egeru and Majaliwa, 2009). Insecurity in the region in the late 1990s as a result of intensified cattle-rustling and armed conflicts involving the Lord's Resistance Army caused mass movements of the population to safer areas, leading to land shortages in Teso (DSOER, 2004). After the war, and after the government had put a stop to cattle-rustling, the wetlands became home to migrant cattle farmers in search of lush pastures in what are considered communal grazing lands (DSOER, 2004; Egeru and Majaliwa, 2009).

The risks of livestock and crop yield losses due to frequent droughts in the recent past have increased, leading to increased farming activities in the lowlands closer to the wetlands. Recent decades have also seen an increase in both the intensity and frequency of flooding, especially around the wetlands, resulting in short-term benefits such as increased fishing activities, water transportation and wetland cropping. Paddy cultivation also seems to have acquired economic importance as a result of this phenomenon (Amaniga *et al.*, 2010; Nabikolo *et al.*, 2013).

All these changes to the Awoja watershed ecosystem have in part been attributed to climate change, commercialization and increased demand for food generated by increases in population (Nature Uganda, 2005). It has also been suggested that high poverty levels, reductions in agricultural land because of increasing populations, climate change and variability and a lack of alternative sources of income, among other things, are likely to have caused increased human activities within the watershed (NEMA, 2004; Egeru and Majaliwa, 2009). Economic activity, including increased commercialization, seems to be compounding flood-related push factors where there is human settlement in flood plains. A further complicating factor is that, due to increased human activity brought about by changes in the demand for commodities, degraded ecosystems are becoming less resilient to climate changes. Thus, the increased demand for commodities has indeed compromised resource protection options in watershed use systems.

Land ownership in the watershed consists of a mosaic of forms of tenure. Farmer's individual plots are typically inherited[1] and although some households do have title deeds, they are far from widespread. Pastoralism in the neighbouring Karamoja sub-region affects livelihood systems in the Teso sub-region as thousands of livestock graze in the Awoja wetlands during the dry seasons. Access to farming land and water is thus a significant point of conflict in the watershed, though fortunately the previously poor road infrastructure is being improved, bringing new interests into the watershed for land, commodities, or tourism and wildlife conservation.

Findings

In accordance with the main hypotheses set out earlier in this chapter, the findings are presented under four different themes: (i) the influence of different political interests on the watershed; (ii) the challenges of institutional multiplicity on management of the watershed; (iii) reinventing legitimacy and relevance through shifting interests; and (iv) innovations, negotiations and compromises in watershed management. The four themes were chosen because they generally reflect the political processes within the watershed, and particularly the interface between the community, local government and civil society and their impact on watershed management. The findings thus indicate that political interests within the watershed are a reflection of the various actors involved and that there are significant effects on the management of the watershed.

Influence of different political interests on watershed management

Different political interests were found to be significant in affecting the management of the Awoja watershed. In this chapter these interests have been divided into two aspects: (i) the different political interests and the conflicts over shared resources; and (ii) limited funding and public dissatisfaction.

The different political interests and conflicts over shared resources

The study found that policy processes in the Awoja watershed are subject to varying levels of political interest. Although local governments are keen to safeguard the sustainability of natural resources through bylaws, rice farmers are looking to increase their use of the wetlands, whereas other community members are interested in forests and sand quarries as alternative sources of income under a stressed socio-economic regime being made worse by a changing climate. It is for this reason that one community member in Amuria lamented: 'the government does not see the needs of the people, that we need food and money to take children to school, and this is why they are forcing us out of the wetland, and small land is being left for rice growing'. A similar feeling was revealed in Angisa parish, where the government had demarcated land for wildlife conservation. A resident of Angara village argued that 'the government does not seem to be bothered by the fact that people are starving, and that people can be chased from their lands to utilize the land for wildlife conservation, in which case wildlife is being substituted to the value of human beings'. In this study, 53% of respondents in the survey stated that the powers to allocate the wetlands should be given to community leaders and landowners rather than the government, whereas a majority of 76% indicated that the powers to allocate the use of forests should be given to government because the forests are not an immediate source of food.

Although natural resources and their protection are held in trust by the government for the people of Uganda (Nature Uganda, 2005; NEMA, 2011), in practice control of natural resources and its ownership is exercised by clan members nearest to those resources, in line with the traditional land-tenure system. It was revealed that those who lived adjacent to resources such as forests and wetlands become the owners of those resources, with the power to use them, rent them out or even sell them to those who expressed an interest in doing so. During focus-group discussions it was revealed that people who live adjacent to the wetlands are the owners and were highly protected as such through clan regulations. The major regulation, as revealed by the Chairperson of Amuria, was that the wetlands belonged to the government but that those who lived next to them were the owners unless the government took over control. During the study, local government officials in Amuria, Soroti and Katakwi also revealed that the traditional guidelines regarding land use had weakened the efficacy and acceptability of bylaws concerned with natural resource protection instituted by local government.

It was also found that decisions over resource conflicts were influenced by the anticipated political implications of decisions, whereby politicians involved in protecting natural resources risk being voted out during elections by the aggrieved communities against which those decisions have been taken. Thus the resolution of conflicts over resource use produces winners and losers and has political repercussions. In the words of the local councillor from Wera subcounty, community members have often promised to meet the politicians in 'the basin'.[2] An example of a politician who was voted out of office due to his poor handling of land conflicts is the Chairman (council level LC3) of Wera, who had been confronted with several challenges during the land demarcation exercise in Wera just before the general elections of 2016. In Angisa in Katakwi, respondents revealed that those elected officials who had not fought for them when the Wild Life Authority claimed and took some of their land as a game reserve will not be voted into power again. The local government leaders also revealed that in many instances they had tried to compromise with the communities because they need them to retain their political positions.

These compromises have also increased struggles in the form of conflicts over wetlands for rice cultivation, as has occurred in Angole Wera, or cutting protected tree species from another person's land, etc. In Angole Wera parish, one community extended its rice fields beyond their village to another village. This created widespread inter-village and inter-clan conflicts over land, as the settlements of a particular clan often cover an entire village. These types of conflict are difficult to solve because of the way the contending parties are interconnected or because of the ethnic dimension that often quickly becomes an issue. The sub-county intervened by demarcating the swamps so that both villages lost the land they were using for growing rice. At election time, both villages voted out their councillors for failing to protect their land and livelihoods.

At the parish level, political influence is evident in the implementation of rules and regulations by the local environment committees who oversee compliance with natural resource regulations, reporting infringements to the sub-county and district authorities. However, in executing their mandates, local environment committees reported most often being unable to ensure the prosecution of encroachers once they have been apprehended due to political inteference. Elite relatives of encroachers exploit the weaknesses in the environmental laws to have the offenders released, thus giving a bad example to others who may have an interest in encroaching into protected areas. It is becoming increasingly clear to the environment committees that their power is weaker than they had initially been made to perceive. Therefore, from the interviews, it appeared that local participation is merely an invited political space devoid of the real power to get encroachers prosecuted. The quantitative results indicated that some offenders were not punished because of compromises by leaders, which among other reasons included administrative errors (25; 13.9%), bribery of leaders/institutions (34; 18.9%), cases not reported (25; 13.9%), friends/relative/connections with the leaders (78; 43.3%), risk of creating enmity (4; 2.2%) and relaxed regulations (12; 6.7%).

During the face-to-face interviews, some respondents indicated that offenders are sometimes reported to the police in the evening and in the morning they are released. This was reported as annoying to the residents and cases are no longer reported. It was further revealed by the residents of Omugenya that some leaders could not punish offenders because they themselves were illegally using the same resources, for example, cultivating rice and grazing in the wetlands. 'They cannot punish those who break the rules because they break them together', said one resident.

According to the Chairman of the Environment Committee of Wera, the power to prosecute encroachers and punish them should be devolved to the committees. Otherwise the committees are frustrated when people they arrest are released without due process. The Teso sub-region has been a hotbed of opposition politics in Uganda, and underlying political confrontations and power relations within the state apparatus play a major role in the activities of technical officers charged with governance of the wetlands. In Amuria district, the technical staff have to tread carefully to balance the interests of the opposition-led district council and the formal mandate of the ruling government. This is not helped by conflicts with other departments, particularly the Department of Agriculture, which officially promotes rice production in the wetlands under JICA, a GOU–Japanese project. The technical staff of the Department of the Environment have to navigate through these and other informal state practices by maintaining their relationships with the central government without compromising their positions in the local institutional context.

The lack of funding and dissatisfaction of aggrieved communities

The study also found that the implementation of regulations protecting natural resources was constrained by limited political support of and funding for enforcement activities. Local government councillors give resource management a low priority in their budgets because protecting resources tends to undermine their political security. In Gweri sub-county, the agricultural extension officer reported that the allocation to natural resource activities was only UGX5 million (Ugandan Shillings) compared to Works being UGX87,100,000, Administration UGX27,399,375 and Water UGX22,100,000. It was revealed that a parish environment officer who is duly elected by all parish members is given a bicycle with which to travel across the whole sub-county and receives only UGX10,000 a month (US$3), far too little to enhance the enforcement of local government rules and regulations over natural resource use. It was also revealed that environmental protection meetings at lower levels of local government are not facilitated financially, reducing the motivation for community members to participate in the meetings.

The study also found that, while 53% of respondents felt that regulations were effective in curbing degradation, 31% felt that the regulations were not being enforced effectively. Discussions with staff of the Department of Environment revealed that the department faces many political challenges. For example, the budgets they command are too small given the vast areas, poor roads, many remote communities and considerable variation in land-use, ecosystems and environmental issues, and the lack of transport. Moreover, the department has a vast mandate, including mainstreaming environmental issues in the district development plan in line with environmental legislation, ensuring that environmental impact assessments are conducted for all projects in the district, enforcing environmental legislation at the community level, supporting the work of the Environment Committee in environmental planning and decision making, providing guidance and awareness to communities on environmental management through participatory approaches, and serving as a contact point for public complaints and grievances under environmental legislation (interview with District Environment Officer, Amuria).

Challenges of institutional multiplicity on management decisions

After the recent droughts and floods, many NGOs have emerged to respond to climate change, thereby changing the institutional landscape in the watershed. Initially these NGOs were concerned with the relief of those displaced by wars, floods, droughts and cattle-rustling. However, informants stated that in the past 20 years their mandates have changed substantially from relief services to development activities and from project modes of operation to greater integration of their activities with local government. The parish chiefs of Omugenya and Wera stated that some NGOs are funding the mainstream activities of local governments, such as joint monitoring, planning and reviews of development projects. They also indicated that NGOs have created working committees attended by both local government leaders and traditional leaders, where deliberations are made on various issues affecting communities, including sustainable adaptation to climate change. In interviews, informants revealed, however, that the relationships between the NGOs and local government show that, as a result of the relatively limited financial and human resources of the latter, some of them seem to be surrendering or delegating their tasks to the former. The local parish chiefs reiterated that the efficacy of NGOs in carrying out some local government functions, such as protecting the environment, is questionable because they cannot enforce environmental protection where this means forcing people to stop using some resources while at the same time enforcing social protection. They further stated that NGOs are more concerned with promoting livelihood options for communities than environmental protection. Examples of such initiatives include the Village Savings and Livelihood Associations (VSLAs) being promoted by World Vision and SOCADIDO (the Soroti Catholic Diocese Integrated Development Organization) in Soroti and Katakwi Districts. SOCADIDO also promotes vegetable production using kitchen gardens as a response to land shortages. Despite the fact that many NGOs do not promote resource conservation effectively, government departments have continued to promote them as alternative modalities for service delivery and to channel funding to community activities. The end result is that the debate on environmental protection remains merely rhetorical, while conservation practices remain weak. Thus resources have continued to be degraded in the presence of NGOs and local governments.

Meanwhile, community-based organizations (CBOs) serve a variety of purposes in this community. While some are merely platforms for access to donor and local government funding, others, such as the Amuria Development Association (ADA), are driven by broader grievances over rights and/or developmental concerns over resource bases, particularly conflicts over land on the boundary between Teso and Karamoja. At the same time, some CBOs in the area, such as the Wera Development Agency (WERDA), are increasingly seeking to assert their authority as managers of the environment in their own right and have established many community tree nurseries. With many NGOs claiming more authority as representatives of the people's interests, they are increasingly becoming legitimate players in the local power game over natural resources.

Local government environmental staff are thus confronted with a context of increasing institutional multiplicity, whereby a growing number of different types of organization are performing functions and claiming authority within natural resource management. It was revealed during the study that the involvement of NGOs is not generally a bad thing because it has yielded benefits. For example, respondents revealed that they have sensitized people to a number of climate change adaptation options and enhanced community involvement in decision making, and that this has generally reduced the level of conflicts over resource use. The parish chiefs revealed, however, that, while the involvement of NGOs provides certain opportunities for local staff and other technical officers to expand their engagement in the implementation of rules and regulations, it also potentially challenges their role as local authorities in the management of the environment because the NGOs tend to take the lead. This contradiction is similar to that highlighted by Ribot *et al.* (2010), where he argues that, when these organizations take over the responsibilities of elected local governments, they run the risk of draining local administrators of authority and resources, thus undermining their legitimacy.

The interaction with sub-county environmental officers in Gweri and Magoro indicated that the role of the Department of Environment in local government is blurred. They revealed that they are supposed to implement NEMA regulations and associated policies as both law enforcers and the facilitators of participatory approaches, and they must do so single-handedly with very limited means across a large area with multiple and competing forms of land use. At the same time, they must navigate a local institutional landscape characterized by informal competition and power struggles within the state itself, as well as a broader institutional environment characterized by growing pressure and expectations from CBOs and challenges to their authority from both civil society and local government. The results also make it clear that central government officials sometimes draw up policies and sit back instead of monitoring them closely to ascertain whether they can be implemented. One environment officer revealed that, in his experience, environmental laws are generally not uniform across the country. He noted that, in interactions with colleagues in other districts, he is surprised that he is actively enforcing the demarcation of wetlands, whereas in other areas nothing is being done. He also observed that wetlands in other districts such as Iganga, Bugiri and Tororo are being indiscriminately used to grow rice, whereas in Amuria they are being demarcated. This makes the implementation of natural resource rules and regulations difficult.

Despite their seemingly weak status, local environmental staff still profess a strong status as law enforcers and as symbols of the state. However, their authority is increasingly being eroded by their multiple accountability obligations.

The study also found out that the traditional role of local government has declined because of resource constraints and the increased integration of NGOs with local government, through which the former take over some of the traditional mandates of the latter, and that CBOs have emerged in response to unfulfilled needs by government arising from resource constraints. During discussions with local government leadership in Wera parish, we were told that some NGOs are funding the mainstream activities of local governments such as joint monitoring, planning and reviews of development projects.

Respondents further revealed that the relationships between NGOs and local government shows that, due to the relatively limited financial and human resources in local governments, some local governments seem to be surrendering or delegating their tasks to NGOs. However, the efficacy of NGOs in carrying out some local government functions, such as protecting the environment, is questionable. Respondents observed that NGOs are more concerned with promoting livelihood options for communities than environmental protection. Yet, through financial support, government departments have promoted NGOs and CBOs as alternative modalities for service delivery and as funding channels for community activities. The end result is that the debate on environmental protection remains merely rhetorical and conservation practices remain weak.

Meanwhile, CBOs serve a variety of purposes in this community. Although some are merely platforms for access to donor and local government funding, others, such as the Amuria Development Association (ADA), are driven by broader grievances over rights and/or developmental concerns about resource bases, particularly contested land on the boundary between Teso and Karamoja. At the same time, some CBOs in the area such as the Wera Development Agency (WERDA) are increasingly seeking to assert their authority as managers of the environment in their own right and have established many community tree nurseries. With many NGOs claiming more authority as representatives of the people's interests, they are increasingly becoming legitimate players in the local power game over natural resources.

Despite their seemingly weak status, local environmental staff still profess a strong status as law enforcers and as symbols of the state. Their authority is, however, increasingly being eroded by their multiple accountability obligations.

Changes in organizational practice

Before the floods of 2007, most NGOs in the three study districts in the watershed (Soroti, Amuria and Katakwi) were working separately on various social programmes. The study found that previously most of the NGOs were focusing on the challenges that existed at the time, such as cattle-rustling by Teso and Karamojong cattle keepers and promoting education and credit and savings institutions. Even then, each NGO was working separately on an individual basis. However, the consequences of climate change, especially those of flooding, have brought most NGOs together because they view the consequences of climate change as a common problem. Informants stated that collaboration between NGOs strengthens their voice when it comes to lobbying for funds and legitimizing their presence in local communities. This development has led, however, to the emergence of new institutional configurations and new rules of engagement among the NGOs involved in addressing the effects of floods, droughts and gradual climate change. The NGOs involved have different focuses, some being involved in sensitization and encouraging the growing of trees and other agricultural practices that can reduce on soil erosion and improve soil fertility. Others are concentrating on promoting proper adaptation mechanisms such as appropriate settlement structures and adopting livelihood enhancement. Irrespective of various orientations, however, most NGOs are focusing on climate change adaptation.

The study found that the post-recovery period following the unprecedented 2007 and 2012 floods that hit the Teso area sparked public concern about environmental degradation and depletion. This event and other drivers (such as occasional prolonged droughts and raids by the Karamojong) have shaped the institutional landscape in the watershed. A constellation of informal agencies has emerged to address the social and economic needs of local communities, including World Vision, the Katakwi Integrated Rural Development project, Katakwi-Amuria Agro-forestry, the Katakwi Rural Community Malaria Project, Katakwi Community Empowerment funded by the Lutheran World Federation, the Climate Proof Disaster Risk Reduction Project funded by TPO Uganda and many other projects by various funding agencies.

A number of institutions are working together to reduce the costs of operations in activities aimed at mitigating the effects of climate change. They work together in organizing meetings and having joint sensitization sessions, and even utilize similar community-based teams to extend messages to the general population. It was even revealed in interviews that other institutions have emerged as community development groups promoting innovative responses to climate change. These are mainly community-based committees rooted in the village level that are mainly concerned with establishing order in resource use. As a result of floods, in addition to demarcating wetlands in some areas, there has been a scramble for resources in the uplands for farming, making it prone to conflicts that have led to the formation of village committees to handle them. They have not replaced the traditional clan leadership committees, although the Omugenya chairpersons revealed that the different stakeholders work hand in hand to ensure that conflicts in resource use are minimized. The interface between the different institutional committees in managing climate change adaptation has provided platforms for the different groups to express themselves and has promoted group-based institutions for resource management. The emerging institutions have led to a number of developments in the process of adapting to climate change, however, including agro-forestry, fuel-saving stoves, creating water channels to drain waterlogged areas, and planning for disaster risk reductions and responses by involving the disaster risk reduction committees at the village, sub-county and district levels that have also emerged (see next paragraph).

The sub-county local governments have now become aware that communities in the watershed are prone to climate change-induced disaster and are particularly looking at measures to sustain community livelihoods. According to the sub-county chiefs in the watershed, the sub-counties have begun to integrate climate change concerns into all their programmes. The sub-counties are creating awareness about the sustainable use of all natural resources within their boundaries. A more proactive approach to disaster responses in the sub-county has been the creation of sub-county Disaster Risk Reduction Committees (DRRCs). The committee develops disaster preparedness and implementation plans that guide formal and informal institutions working in the sub-county. These plans also set out what needs to be done before the disasters occur in order to make communities more resilient. The sub-county DRRC is composed of political leaders, sub-county technical staff (the Senior Assistant Secretary as chair, the CDO and the National Agricultural Advisory Services coordinator) and a representative of each development partner in the sub-county (Lutheran World Federation, Trans-cultural Psychosocial Organisation-TPO Uganda, and Family Life Survival-FALISU).

The emergence of different institutions has eased the roles of local governments in various ways. It was reported that the emerging NGOs have helped in the implementation of central government environmental guidelines. Furthermore, the Omugenya and Wera parish chiefs reported that the NGOs are helping to sensitise community members to respect the government's rules. These rules include guidelines governing demarcation and encouraging parents to take their children to school in order to utilize the opportunity presented by the introduction of universal primary education and universal secondary education. Hence the NGOs have tried to support the sub-county and parish environmental monitoring teams by being closer to the population. The changing roles of institutions have contributed to the process of adaptation to climate change, as there are now many stakeholders involved in aiding the process.

In Angisa parish, however, it was found that the shift in institutional interests was different. As a result of climate change the roads had become impassable, and this has led to a number of institutional changes. It was revealed that NGOs have shifted their operations from rural Angisa to more accessible areas like Magoro. Rooted in this, one respondent noted that 'Even NGOs have left us alone as a result of floods as roads have been blocked and bridges destroyed'. The clan leaders had become stronger since most of the elected leaders do not live in the village. Residents revealed that after the elections, the elected leaders started feeling their power and most of them relocate from the rural areas to towns because they think they are now a class apart. This has left the responsibility for addressing community challenges in the hands of clan leaders even when a local government structure exists. This form of institutional interest has disadvantaged community members because some of their challenges needed immediate attention from local government leaders, who, however, remove themselves from access. This change was confirmed by the Chairman of Angara village when he said the reason the elected leaders are moving to town is to make it easier for them to provide services to the people. Community members noted in interviews that this was not true, however, and that the elected leaders were simply running away from bad village conditions. The Chairman of Angara village further explained that it is not easy to travel from Angisa to the sub-county, especially during the rainy season, yet they must attend meetings at the sub-county. As a result, they have left the village to stay near the sub-county, though community members are not happy about it.

Innovations, negotiations and compromises in watershed management

The political situation in the watershed is characterized by innovations that involve the process of discovering new ways of co-existence within the watershed between climate change and undesirable rules and regulations. Negotiation is a process of community members agreeing with law enforcement officers that local government rules and regulations should be changed to fit people's interests in the context of adapting livelihoods. Compromise refers to a situation in which community members make concessions in a negotiation over a common good. Conversely, inaction and constructive ambiguity mean allowing the degradation of the watershed while the leaders are watching and failing to act because of the fear that they might break their relations with community members. All these processes have affected the management of the watershed in one way or another, as the preceding account has indicated.

In the case of the negotiations, it was revealed that where demarcations have succeeded, staff from the Department of Environment have been persuaded by communities to neglect implementation of the natural resource management regulations in order to create harmony with the communities. In the Amuria wetlands, the authorities negotiated with the communities to reduce the official buffer zones from 30 m to 15 m from the centre of the wetlands to allow people to grow rice there, as well as some rapidly maturing crops to avert the threat of famine. These negotiations are often made at village-level meetings. Community members use convincing arguments, such as the problem of famine during drought or the need to raise money for health-related reasons, school fees, etc., to renegotiate the rules. Increasingly, leaders have allocated parts of the wetlands for the emergency production of vegetables for food security despite regulations against using the wetlands. This has only served to legitimize the use of wetlands for crop production. Successful renegotiation of the rules is a big political benefit for community members but it only serves to legitimize improper use of the wetlands. According to the environment officers, however, these types of concessions are only tactical in that they enable wetland demarcation to be peacefully introduced. They hope to renegotiate with communities to adjust the buffer zone back to 30 m once communities appreciate the need for buffer zones.

The aspect of innovation was also identified during the study and has also turned into a political issue because it has created an interface between local government officials, NGOs and community members in a process of aiding adaptation. For example, it was revealed that, although the renegotiation of restrictions on using the wetlands for crop production have been driven by the dire need for communities to survive during droughts, in Magoro sub-county in Katakwi District the renegotiation of the rules has arisen from innovations. One of the key effects of floods has been the collapse of huts built from unburnt mud bricks, which led communities to demand the rules against cutting down trees be waived to allow people to use the poles to build mud and wattle huts that are more resistant to floods. NGOs working with the Disaster Risk Reduction Committees at the sub-county, parish and village levels developed a technology, however, for building flood-resistant huts using a damp-proof course made of polythene paper for the floor. This cuts off the floodwaters that would dissolve the mud blocks and result in the walls collapsing. Once this innovation had been generally popularized, the councils reintroduced the rules against cutting down trees to protect the environment.

Another example is of communities using receding floodwaters to grow paddy rice. Paddy rice is grown in the uplands using receding floodwaters and is technically not subjected to the rules governing rice growing in the swamps. As a result of this innovation, communities are able to dodge the rules altogether.

In yet another case, in Magoro, in a bid to reduce the effects of floods, local communities agreed to construct water-drainage channels through complex social and ecological terrains. This has helped to direct the floodwaters away from houses and gardens to nearby valleys. This has been a great innovation and has involved a political process because community members must first agree where the waters should be drained and whose plot of land will be affected. The conflicts arising from this innovation are therefore resolved by community members themselves in cooperation with the local leadership.

Discussion

Following the presentation of these findings, the following discussion will reflect on the contextual and theoretical perspectives of the study. The discussion has helped clarify the aspects of power and accountability mechanisms, the tendency of politicians to use their power to allocate resources, and the tendencies of NGOs to account upwards to their donors and not be bound by conservation rules. This also shows how local staff are caught between a local power game and national obligations to initiate institutional change in response to community needs (bricolage).

Owing to the current climate change that is affecting the world, wetland utilization has been acknowledged as a major social and economic challenge for nations, and most of the emerging policies have tended to limit access and utilization rights because of environmental concerns (Adger *et al.*, 2001). These initiatives have aimed at enhancing alternative and adaptive capabilities for those who are highly dependent on natural resources (Lund and Treue, 2008). However, policy deliberations are a bargaining process over social power, where representation and bargaining power determine both the winners and those whose interests are compromised (Rantala, 2012). In Uganda natural resources are being kept in trust for the people by governments, whereas participatory policy and enacted laws result in resource guardianship. As can be seen in this study, the demarcation of wetlands ensures that communities are excluded from using certain resources to ensure sustainability. However, where exclusion becomes a major protection approach, communities that might have lived in harmony with resources for many years could be viewed to be acting detrimentally to these same resources (Logan, 2004). It should be noted that in most African countries, a lack of funds and human resources is leading to governments failing to maintain exclusion practices or to manage resources effectively (Wily, 2000). This may lead to competition among interest groups and institutions and weaken the legitimacy of adaptation measures.

Divergent political interests are a manifestation of the dilemmas facing policy and the implementation of natural resource regulations. The management of natural resources has traditionally been framed around the idea of the 'state'. Public servants who represent the 'local state' potentially have had significant influence on the options and limitations for community management of collective resources and are seen as an embodiment of autocratic state-led approaches (Keeley and Scoones, 2000). However, in framing policies, state agencies are increasingly expected to act as 'service providers' to the people, while their authority as representatives of 'the public' is made subject to competition (Lund and Treue, 2008). The study shows that public servants must relate to and position themselves with regard to local resource struggles both within the state bureaucracy itself and in local politics (Bierschenk, 2014). In this way, local policy structures are subject to being compromised within the bureaucratic system, as demonstrated by the relaxation of environmental regulations in Amuria.

Resource management in a changing climate has also seen the emergence of many NGOs and CBOs in local communities. Thus, while public services for natural resource management have traditionally been the responsibility of the traditional 'state', the state is no longer the sole actor in collective service provision, there now being a multitude of formal and informal actors and institutions in the field of public authority (Neumann, 2005). Therefore, although bureaucratic authority defines the state, there is a recognition that local governance is more than just a clear-cut question of official bureaucratic authority. It involves a complex process of negotiation between various actors and groups referred to as 'real governance' (Olivier de Sardan, 2011).

One characteristic of the political process in watershed management is institutional and technical innovations. According to Chhetri *et al.* (2012) climate change may alter the environment. Such changes provide appropriate signals to farmers and their supporting institutions to induce technologies suitable for the new environment. The study further shows how induced technological innovations such as use of receding floodwaters for rice growing and the construction of flood-proof huts have been used by local governments as well as by communities to renegotiate aspects of natural resource management rules.

Conclusions

The governance of natural resources in the Awoja watershed is subject to the different political interests of multiple stakeholders. The decentralized governance of climate change adaptation coupled with the limited financial resources available at the local government level has provided a space for the participation of many formal and informal actors. This has weakened the power relations for the enforcement of natural resource rules. Participatory governance is constrained by the challenges of multiple accountability, where local governance staff need to account upwards to central government and downwards to the community, as well as sideways to their relatives and kin. This has far-reaching implications for the effective management of resources in the watershed. Stakeholder participation in management committees has created a space for innovations and the renegotiation of rules and regulations for the governance of natural resources under extreme climatic conditions. NGOs and CBOs have initiated many innovative measures for the management of the watershed. In this way, their legitimacy as advocates of sustainable resources has increased at the expense of local government regulatory authorities.

Notes

[1] This is where land is transferred from father to child, which in the Teso culture means its transfer mainly to sons. Daughters are not expected to inherit land, and they can only use it. Source: field results, 2016.

[2] That is, an election booth.

References

Adger, N.W., Benjaminsen, A.T., Brown, K. and Svarstad, H. (2001) Advancing a political ecology of global environmental discourses. *Development and Change* 32, 681–715.

Amaniga, I.R., Lucy, L., Mafumbo, J., Nabulumbi, J., Mwesigye, J. and Madanda S. (2010) *A Socio Economic Baseline Survey of Communities Adjacent to Lake Bisina/Opeta and Lake Mburo/Nakivali Wetlands System: Providing Information for the Implementation of the Cobweb Project in Western and Eastern Uganda*. Uganda Wildlife Authority.

Bierschenk, T. (2014) Introduction. In: Bierschenk, T. and Olivier de Sardan, J.-P. (eds) *States at Work. Dynamics of African Bureaucracies*. Brill, Leiden, The Netherlands.

Chhetri, N., Chaudhary, P., Tiwari, P.R. and Yadaw, R.B. (2012) Institutional and technological innovation: understanding agricultural adaptation to climate change in Nepal. *Applied Geography* 33, 142–150.

DSOER (2004) Soroti District State Of Environment Report 2004, Soroti District Council.

Egeru, A. and Majaliwa, M.G.J. (2009) Landuse/cover change trend in Soroti District, Eastern Uganda. *Journal of Applied Science and Environmental Management* 13, 77–79.

Keeley, J. and Scoones, I. (2000) Knowledge, power and politics: the environmental policy-making process in Ethiopia. *The Journal of Modern African Studies* 38, 89–120.

Logan, B.I. (2004) *African Environment and Development: Rhetoric, Programs, Realities.* Ashgate Ltd, Aldershot, UK.

Lund, J.F. and Treue, T. (2008) Are we getting there? Evidence of decentralized forest management from the Tanzanian Miombo Woodlands. *World Development* 36, 2780–2800.

Mwangi, E. and Wardell, A. (2012) Multi-level governance of forest resources. *International Journal of the Commons* 7, 339–343.

MWE (2007) *The Water & Sanitation Sector Performance Report*. District State of Environment Report, Kampala, Uganda.

Nabikolo, D., Bashaasha, B., Mangheni, M.N. and Majaliwa G.M. (2013) Determinants of climate change adaptation among male and female headed farm households in Eastern Uganda. *African Crop Science Journal* 20, 203–212.

Nature Uganda (2005) Ramsar Information Sheet (RIS), Wetlands International. Available at: jncc.defra.gov.uk/pdf/RIS/UK11081.pdf (accessed 17 January 2017).

NEMA (2004) *Soroti District State of the Environment Report (2004)*. National Environmental Management Authority (NEMA), Republic of Uganda.

NEMA (2010) *Annual Performance Report 2010/2011*. National Environmental Management Authority, Kampala, Uganda.

NEMA (2011) *National Environmental Compliance & Enforcement Report*. National Environmental Management Authority, Kampala, Uganda.

Neumann, R.P. (2005) *Making Political Ecology*. Hodder, Arnold, New York.

Olivier de Sardan, J.-P. (2011) Auth-rent States, bureaucratic mode of governance and practical norms, in West Africa and beyond. Contribution to the conference on *Governance Beyond the Centre. Informality, Institutions and Contested Power Structures in Authoritarian Contexts*, Berlin.

Rantala, S. (2012) Knowledge and brokerage in REDD+ policy making: a policy networks analysis of the case of Tanzania. *Sustainability Science Program Working Paper No. 2012-03*. Sustainability Science Program, Kennedy School of Government, Harvard University, Cambridge, MA and Center for International Forestry Research (CIFOR), Bogor, Indonesia.

Ribot, J.C., Lund, J.F. and Treue, T. (2010) Democratic decentralization in sub-Saharan Africa: its contribution to forest management, livelihoods, and enfranchisement. *Environmental Conservation* 37, 35–44.

Wily, L. (2000) *Making Woodland Management More Democratic: Cases from Eastern and Southern Africa*, Gatekeeper Series. IIED, London.

Image courtesy of Tine Engedal

10 Competing Interests over Natural Resources and Adaptation to Climate Change: the Case of Rice Cultivation in the Gweri Wetlands, Uganda

Charles Aben, Esbern Friis-Hansen, Bernard Bashaasha and J.J. Okiror

Introduction

Nationally, responses to floods and droughts have generally been 'reactive, poorly coordinated and untimely' (Wilhite and Pulwarty, 2005). As part of the decentralization reform in Uganda, the formal governance of national resources was devolved from the Ministry of Water and Environment to the Department of Natural Resources at district level. As an effect of this devolution of power, along with relevant legal frameworks for the protection of wetlands, local governments were expected to offer more relevant and timely support for more sustainable natural resource management, including flooding associated with extreme climate change events. The devolution of responsibilities was not followed, however, by the adequate provision of fiscal resources. As a consequence, the Department of Natural Resources and its institutions at the community level, the Sub-county and Parish Environmental Committees, have remained weak, unable to enforce regulations and subject to local elite capture.

In the absence of strong government institutions, local institutions compete for authority over how to interpret traditional land tenure laws and to enforce local government regulations. Such local institutions include the traditional clan structures and enterprise associations, for example, of cattle herders. In the Gweri wetlands, the focus of this study, the institutional arrangements to manage the wetlands are competing and giving rise to contestations over the interpretation of land-use management regulations. The competing institutions are a reflection of a wider conflict between different socio-economic constellations and environmental forces within the community.

A decade ago, cattle herders were using the Gweri wetlands as a grazing area in peaceful co-existence with rice growers, fishers and other traditional users of the wetland's resources. Recently, however, social conflicts have erupted. Whereas historically there has been some competition over these resources among the different actors in the Awoja wetland system, the competition in the Gweri wetlands has become fiercer during the past decade with the changing climatic conditions and the advent of multiple interest groups.

Interviews among key informants living adjacent to the Gweri wetlands revealed two major trends that are responsible for the increased tension between different land-use groups during the past decade: climate change and the commercialization of rice cultivation.

The first trend is *increased climate change hazards* or, more specifically, the effects of a combination of gradual and extreme climate change. Gradual climate change has lowered the water levels along the banks of rivers and swamps, increasing conflicts over water between people involved with different enterprises. The fishers can no longer keep the deeper waters to themselves, whereas cattle herders who for generations have left their cattle to roam the wetlands freely for pasture find that their grazing areas have been taken over by crops. This restricts the movement of cattle, causing livestock herding to become more labour intensive and increasing livestock encroachment on cultivated fields. Gradual climate change has also caused farmers living adjacent to the wetlands to expand their cultivation into that area to take advantage of its high soil moisture, a reaction to the disruption to traditional agricultural seasons because of the increased variability of rainfall.

Extreme flooding affected the Gweri wetlands in 2004, 2007 and 2010. All stakeholders agree that the wetlands played a crucial role during the extreme floods as sinks for the excess water. Without the wetland swamps, the consequences of extreme flooding would have been much more severe. Regulation of the wetlands through sustainable land-use practices that preserve the wetland status of the area can be seen as a collective adaptation to floods.

The second trend is the *expansion of the commercial tractor cultivation of rice*. Well-off interest groups located outside the community in collaboration with commercial rice growers are clearing large areas of the wetlands for cultivation. They are also claiming ownership of the wetlands, which for generations were considered to be communal land that belonged to entire communities on which no single individual could claim tenure rights. This value chain-driven expansion of rice cultivation is fuelling two major land-use conflicts. First, the expansion of commercial rice production is degrading the wetlands and slowly turning them into dry land. This process has a range of ecological effects, including compromising the ecological sustainability of this important Ramsar site, in effect reducing a range of essential environmental services, including water supply for a large upland area including Soroti town, as well as negatively affecting biodiversity and wildlife. An equally dramatic effect is that expanding commercial rice cultivation reduces the wetlands' capacity to act as a sink for extreme floods and thus increases the area affected.

The expansion of commercial rice cultivation is also creating land-use conflicts with cattle herders, who for generations relied on the wetlands as communal grazing grounds where cattle roamed without restriction. Conflicts arise between cattle herders and commercial rice growers because the traditional free-range cattle grazing is incompatible with the advent of rice and horticultural crop cultivators.

Study focus

Local-level government bureaucrats are divided between enforcing environmental policies and local bylaws that seek to protect both the wetlands and the commercial interests of the local elite involved in commercial rice cultivation in them. The weakness of local government institutions that are responsible for the conservation of wetlands is revealed by their being compromised by interest groups involved in commercial activities in the wetlands. Yet, there is awareness that the conservation of wetlands is important if the issue of adaptation to extreme climate change events such as floods and droughts is to be addressed.

This research aims to examine the interactions of different economic interests in the management of the Gweri wetlands. The overall objective is to understand the interactions of institutions and enterprises and their consequences for climate change adaptation in the Awoja wetland system.

The empirical questions we ask include: Who owns the wetlands? Who are the rice growers? What roles are being played by business people, local government professional staff, and the political and economic elites of local communities in controlling rice production in this wetland? Are small-scale farmers also involved in rice production? Are rice producers just following the water during drought conditions as an adaptation measure?

The following hypotheses are proposed:

- Hypothesis 1: The emerging market for rice is the key driver in the degradation of the Awoja wetland system.
- Hypothesis 2: Elite capture of weak local public wetland management institutions has accelerated degradation of the Awoja wetland system.

- Hypothesis 3: The conflicts over wetlands resources are being exacerbated by fuzzy land property rights contested by the various interest groups and weak local governance institutions.
- Hypothesis 4: Competition between various local public and quasi-public institutions has contributed to the degradation of the Gweri wetland systems.

Methodology

A case-study design was used to obtain a detailed understanding of the competing interests surrounding the use of the Awoja wetlands in Soroti District. Data were collected for a period of 6 months in 2015–2016 using key informant interviews with local leaders in Gweri sub-county and focus group discussions with groups of farmers to improve understanding of and validate the issues identified during the interviews.

This was preceded by a regional stakeholders' consultation workshop in Soroti that brought together 60 participants from nine district governments whose policy decisions and actions impact on the Awoja wetland system in one way or another. The districts that participated in the workshop were Soroti, Katakwi, Amuria, Ngora, Serere, Kumi, Napak, Nakapiriprit, Moroto and Mbale.

In addition, a historical trends analysis of land-use practices in the Awoja wetlands was used to understand the recent changes due to climate change and commercial farming interests. The analysis considered the period before and after the 1980s, when the government introduced rice as a high-value crop in the Teso region under the Rural Farmers' Scheme, as well as the 2007 flash floods on the Gweri wetlands, which were widely attributed to climate change in the area.

Empirical context

The Gweri wetlands are part of the Awoja wetlands system, which covers 14 districts in Eastern Uganda. The Awoja wetland system consists of a confluence of rivers from the vast catchment area stretching from the slopes of Mount Elgon to the East and the mountain ranges of Napak, Toror and Moroto in the Karamoja sub-region. This is the convergence point that links Lakes Bisina and Opeta in Kumi and Katakwi districts to Lake Kyoga in the Lango sub-region in Central Uganda, eventually feeding into the River Nile system. The catchment area is formed by raised mountain ranges on the Kenya–Uganda border that descend into the Lake Kyoga basin in Central Uganda.

The Awoja system is best remembered for the widespread flooding of 2007, which affected the road infrastructure, food security and livelihoods through the mass displacement of people in the Teso sub-region. This wetland system has been the major source of fish in the region and of water for major towns such as Soroti, Ngora and Serere. It has also been used for grazing cattle, firewood and, more recently, rice growing. The fluctuating water levels in recent years, arising from increased torrential rains in Karamoja and Mt Elgon followed by prolonged droughts, have exacerbated the conflicts over resources in the Awoja commons.

Theoretical and Conceptual Framework

The competition for resources in the Awoja wetlands can be understood from the perspectives of political ecology and path dependency. Political ecology allows us to understand the effects of policies and market conditions from local people's perspectives. The theory is helpful in understanding the political forces at work in the struggle for wetland resources and the coping strategies adopted at the individual, household and community levels as a response to the resulting changes (Robbins, 2004).

Society's interaction with land-based resources is linked to local, regional and global power structures. According to Hornborg *et al.* (2012), power asymmetries and social inequalities shape patterns of land use and the management of natural resources. Political ecology is therefore a useful approach to understanding the underlying processes of problems such as environmental degradation and marginalization because it combines the concerns of ecology with a broadly defined political economy such as the equality of cattle herders, rice growers and other resource users on the Gweri wetlands.

Governance institutions and users of wetlands have vested interests that are linked together, but have different functions, powers and relations of accountability that depend on historical socio-political structures. Actors can be differentiated on the basis of their ideology, wealth, objectives, the internal structure of their organization, membership, funding sources and the laws to which they are subjected. Characterizing actors is therefore important in understanding socio-ecological dynamics, especially when the interests of different levels of actors in pursuit of survival contrast greatly.

In the Gweri wetlands, the governance of resources is being constrained by the dilemma of conservation on one hand, and allowing the resource to be exploited for commercial gains in the face of extreme climate change pressures on the other. Moreover, the bureaucrats and the political elite who should enforce the conservation by-laws depend on the local resource users for their survival in electoral politics. As a result, there is laxity in the enforcement of by-laws to the detriment of the ecology of the wetlands.

Continued rice production in the Gweri wetlands can also be understood from the theoretical perspectives of path dependency (Campbell, 2010), where local governance institutions are reluctant to implement natural resource rules adequately owing to vested interests (Fig. 10.1). Actors have taken advantage of this to connive in breaking the regulations together with local governance people charged with enforcing environmental rules, and to alter rules for their own convenience.

Findings

The emerging market for rice as a key driver of competition in the Awoja wetlands

The Awoja wetlands are located near to Soroti municipality, along the transport corridor from Mombasa to South Sudan. Strong market demand has emerged for rice in South Sudan following new trading opportunities with the Sudanese after their recent independence. The respondents in Soroti market explained that the coming of peace to South Sudan has led to the expansion of trade for Uganda. It was reported that the emergence of the market in South Sudan has boosted the trade in cereals, especially beans, maize and groundnuts. It was also reported, however, that the traders in South Sudan do not have the same opportunities as the people of Uganda because the same cereals are also produced in Uganda, and South Sudan is just gaining in strength economically, socially and politically. It was also reported that the high level of market demand has led to a flow of soft loans from millers and traders. This was associated

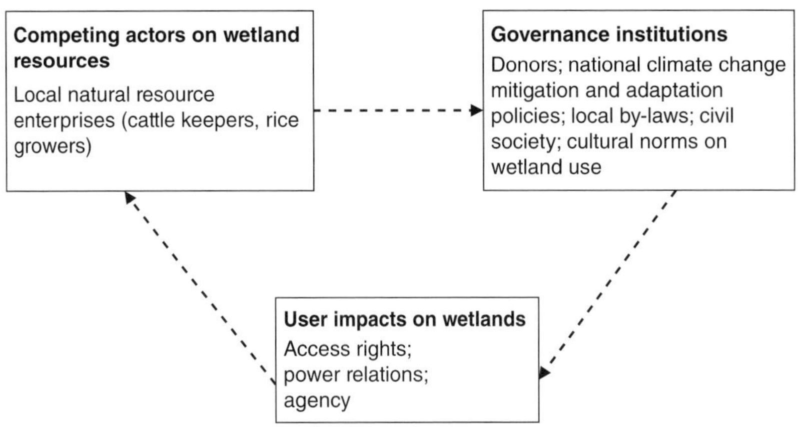

Fig. 10.1. Conceptual framework for understanding competing institutions over wetland use. Source: adapted from Chhetri *et al.* (2012).

with a situation in which the rice farmer can receive money from the trader after the latter inspects the rice garden. Just after the trader has inspected the rice and observes that it is growing in the garden at whatever stage, the rice farmer starts negotiating over payment. It was also revealed, however, that this had become a source of conflict over resource use, especially where the farmer receives loans from more than one trader by promising that he or she will sell the rice to him when it ripens. Problems were reported with farmers still selling the rice to other traders, something that had become a big challenge for local government officials seeking to harmonize the relationship between the traders and the farmers. The area has also seen the growth of contract farming arrangements along the rice value chain. This has been boosted by the government's promotion of upland rice through JICA, a Japanese agency that has been distributing seed to farmers and demonstrating how to grow upland rice as well as paddy rice. Upland rice is not only being grown upland, however, as reported by rice farmers on the fringes of the Gweri swamp. In other words, rice is being grown both in the wetlands and on the fringes of the swamp, thereby increasing the exposure of river banks to erosion and worsening the problem of their degradation. Moreover, upland rice has a higher level of productivity and a similar aroma to the paddy rice or 'super rice', while having characteristics similar to lowland rice in terms of grain size and colour. This allows the rice mills and traders to blend upland rice grown in the wetlands with lowland paddy rice and to sell it as the premium rice of pure paddy. From discussions with participants, however, the major challenge mentioned was that the farmers in Gweri had not embraced the upland rice because of the poor soil fertility of the uplands, which does not favour the growth of rice. As a result, community members have continued to encroach on the wetlands to grow rice ever since demarcation.

During discussions with respondents, it was further revealed that, as a result of the end of the civil wars that have been disturbing the area and the control of cattle-rustling, people have settled down in their communities. This has increased the population in the area and has actually increased the market for rice and the need for rice production. It was reported that people look on rice production as the best means of obtaining money to be spent on school fees and family welfare. This has therefore increased the need for land, from which the rice can be cultivated to meet these needs.

From our discussions with participants in Omugenya village, it was revealed that currently there are more institutions in the area, especially schools, than in the past. This has increased the need for food to feed the pupils in these schools. The main food used to feed them is basically maize, beans and rice. This has greatly increased the demand for rice, as well as for land needed for rice production.

Finally, it was revealed that, as a result of drought in the neighbouring areas, especially the Karamoja region, the population has increased within the region. It was reported that most of the in-migrants are interested in buying rice to send back to their home areas. This has increased the demand for rice in the area and increased the need for land.

Elite capture of local public wetland management institutions as an accelerator of competition leading to degradation

The study uncovered poor implementation of existing natural resource management policies and regulations owing to the elite capture of local government institutions. For example, members of local government committees are also directly involved in cultivating rice in the wetlands. It was found that non-poor farmers and the local economic elite mostly pursue rice cultivation. According to a local opinion leader, one of the reasons why local government institutions do not enforce environmental regulations is that they receive bribes from well-off rice cultivators to gain access to the wetlands.

Another complicating factor is political patronage, where local leaders protect some political supporters and fear taking action so as not to lose the support of the economic elite during the next elections. The lack of the political, human and financial resources to support local natural resource management organs has been another constraint on policy implementation.

Fuzzy land property rights as a contributor to the degradation of the Awoja wetlands system

Participants in the study revealed that wetland stakeholders, including rice growers, cattle herders, fishmongers, charcoal burners, women collecting firewood and thatching materials and others, claim varying forms of ownership and user rights. Moreover, there are also conflicts over ownership rights between cultural leaders and the formal public system. It emerged that the cultural leaders have strong powers over the control and distribution of land resources. It was reported, for example, that, despite the existence of public laws on sharing land amongst families, the people of Soroti, Katakwi and Amuria were applying their own traditional land regulations, such as not giving a child full rights over a plot of land until he grows into a man, until when the land remains in the custody of the clan leaders. It was also revealed that women were not given land, especially if they are married, though when they divorce and return home, they are given land to cultivate without being able to sell it or rent it out to other people. This aspect of land ownership greatly affected land use because, even when a woman had the capacity to use the land, she still faced restrictions.

Concerning the ownership of communal land, especially the wetlands, it was revealed that those who lived near the wetlands were considered the owners of them and could use it as much as they wished. This situation existed despite the existence of government laws concerning the use of wetlands. This has implications for who can access the wetlands for rice cultivation because those living near them prevented other users, especially cattle herders, from accessing them. It also emerged, however, that in areas where demarcation had taken place, there was equal access to all members of the community except cultivators. Fuzziness does not necessarily mean that users do not understand the existing agreements, only that they can challenge the status quo with the intention of creating a new equilibrium, of which the elites can take advantage. Fuzziness as a weakness in local government institutions has led to the stalling of the demarcation of borders between the cultivated land and wetlands, thereby leaving the latter as a free-for-all system with extensive production of rice until the soils were completely depleted. Our research found that in Omugenya Parish in Gweri sub-county, a massive competitive livestock-raising system where all types of livestock are mixed in one wetland is being practised following the decline in rice production. It was revealed that, although there had been no demarcation, farmers had abandoned the wetland to raise animals. This has created a community free ranch where everybody has access to the wetlands for this purpose. From this it can be deduced that where there is demarcation there is free access, but only for raising livestock, whereas access to the upland is still limited owing to the existence of family land boundaries monitored by the clan leaders.

Competition among various local public and quasi-public institutions have contributed to degradation of the Awoja wetland system

The various institutions that manage wetlands in the watershed have different interests, which, as our findings show, have not been harmonized with the capacity of the wetland. Many non-governmental organizations (NGOs) came together after the 2007 floods to offer relief assistance to flood victims. Now they have graduated to tackling development and income generation. Although formally coordinated by the Disaster Risk Reduction (DRR) Committee under local government auspices, coordination has declined with time, and many are now pursuing activities that are not necessarily in line with wetland conservation. The DRR Committee is charged with planning for disaster response at the sub-county and district levels. Although the committee is duplicated at the parish level to oversee compliance with environmental rules and regulations, it faces an uphill task, as well as challenges of legitimacy arising from unclear rules and regulations. They are also subject to intense economic interests from commercial rice growers and millers and are often easily compromised. The findings of this study revealed that environmental rules and regulations were not being properly implemented. The data also revealed conflicts of interests at all levels where, for instance, community members felt that the government should retain the power to control the forests, while the wetlands should be left in the hands of traditional or community leaders. This would create more disastrous ecological consequences, as demonstrated by wetland use in Omugenya Parish, where the wetlands were competitively used for rice production until the soils had been fully exhausted.

Discussion

This chapter has examined competing land-use interests in the Gweri wetlands system as part of the broader Awoja watershed in the context of the changing climate. The study's findings show that the sustainability of the Gweri wetlands is being compromised by the failure of local government officials and institutions to implement environmental laws. Ideally, local council monitors the wetlands through parish environment committees. These committees operate in an ad hoc manner, however, without a specific plan or a budget from the local government. Members of the wetland committee see themselves facing many challenges, including resistance from local communities owing to a lack of sensitization, their own limited technical capacity and a lack of transport provision with which to monitor the wetlands.

Blundo (2015) provides a detailed analysis of 'multiple accountability', a dilemma faced by local leaders when implementing laws among their relatives in circumstances where they have to account to the formal authorities as well as to their relatives, who elect them to these positions. Recent studies suggest, however, that, if an official is solidly 'embedded' in the local social networks, he will feel a strong moral obligation to act for the collective interest (Tsai, 2007). The example of local committees in Gweri failing to implement the regulations could be understood from this perspective.

Our findings also underscore the effects of the rising pressure to increase rice cultivation in the wetlands, initially as a result of upland rice introduced by the government, the private sector, JICA and other non-government agencies. Among the many drivers of rice production is its potential to generate high levels of income for communities. The introduction of upland rice has not helped to draw rice growers away from the swamp because in Gweri the uplands are not very fertile or sufficiently moist for rice production owing to the sandy-loam nature of the soils, which drain rapidly. Most so-called upland rice is therefore grown in the lowlands adjacent to the wetlands and there are not many alternative enterprise opportunities that bring in an income like rice. In the early 1980s, rice was introduced by the Rural Farmers' Scheme (RFS) of the Uganda Commercial Bank and instituted countrywide as a non-traditional cash crop that could finance rural farming communities. Rice is not profitable when produced on a small scale so the Uganda Commercial Bank introduced tractors in the 1980s, setting a precedent that caused massive silting of the Awoja river, the development of floating islands and massive reductions in water levels. Moreover, this tractorization blended well with Teso farming systems that rely on ox-cultivation.

The introduction of rice production by JICA, a government-supported programme, has created some confusion locally over the government's seriousness about its own conservation agendas. Introducing upland rice to pull people away from the swamp has not worked because the uplands are too dry and infertile to grow rice. Upland rice is now growing on the fringes of the swamp and worsening the silting of the wetlands. The proximity of Awoja to the main urban area of Soroti exposes it to pressure from the growing urban population for more rice production. Businessmen are funding rice production to feed their rice processors and the towns. Rice is a short crop that attracts many producers.

Regarding the resource conflicts on the Gweri wetlands, we argue that bureaucratic arrangements for managing conflicts in a watershed will only work if local governments and communities embrace the legitimacy of the rules and regulations. However, given conflicts over the ownership of the wetlands and the failure to demarcate them, local government at the sub-county level has not proved effective in solving conflicts among resource users in the watershed. This has led to the emergence of extra-local organizations and ad hoc systems for conflict resolution. The main conflict among the wetland users is the tension between rice growers and cattle herders over land rights. Conflicts have increased because of the commercialization of rice and the rising numbers of cattle. The conflicts are fuelled by the communal attitude to user rights and a lack of knowledge about government ownership of the wetlands. The conflict is more pronounced in areas where there is limited space left for cattle trails. Such conflicts are either amicably settled by the affected parties or referred to the village local councils for arbitration.

Land disputes also exist amongst rice cultivators because there are no clear boundaries in wetland areas owing to the communal tradition of land tenure. Some farmers claim that wetlands are 'God-given', while others think they belong to them. The majority of the farmers interviewed for this study think that no one owns the wetlands. In many cases, conflicts are resolved through community dialogue, sensitization and the enforcement of wetland management laws by the District Environment Office, sub-county governments and NGOs, which sensitize the wetland users. The framework for the larger Awoja wetland system has been developed and implemented by JICA.

This study has also highlighted both the emergence and the increased role being played by informal alternative institutions. The emergence of the committee headed by the chairman known locally as 'Mr Badman' to manage the wetlands was a decision by the Resident District Commissioner (RDC) of Soroti in response to the failure of the sub-county wetland management committee to curb encroachment on to the Awoja wetlands. Informal organizations enjoy legitimacy with the local communities and are energized to do work that could have been done by the government. The committee was duplicated in the neighbouring districts, being fuelled by the network of RDCs in the region. Its effectiveness has declined, however, owing to the lack of support from the sub-county following the transfer of the RDC who instituted it.

Conclusion

Competition, elite capture, emerging markets for rice and fuzzy laws are all compromising wetlands conservation and undermining climate change adaptation efforts. The study also shows that institutional actors have vested interests which undermine efforts to conserve the environment and erode their legitimacy in managing the wetlands in the public interest. Thus local government institutions have been weakened by commercial interests, combined with the dilemmas of multiple accountability (Blundo, 2015), where officials at the sub-county local government level are caught between accounting for themselves upwards to the district and sideways to their communities.

The key lesson learnt is that any future introduction of a new enterprise needs to take into consideration the ecological implications of the enterprise and the social implications of resource use arising from it.

References

Blundo, G. (2015) The King is not a Kinsman: Multiple accountabilities and practical norms in West African Bureaucracies. In: de Herdt, T. and Olivier de Sardan, J.-P. (eds) *Real Governance and Practical Norms in Sub-Saharan Africa.The Game of the Rules*. Routledge, Abingdon, UK, pp. 142–159.

Campbell, J.L. (2010) Institutional reproduction and change. In: Morgan, G. (ed.) *The Oxford Handbook of Comparative Institutional Analysis*. Oxford University Press, Oxford, UK.

Chhetri, N., Chaudhary, P., Tiwari, P.R., Yadaw, R.B. (2012) Institutional and technological innovation: Understanding agricultural adaptation to climate change in Nepal. *Applied Geography* 33, 142–150.

Hornborg, A., Clark, B., Hermele, K. (2012) *Ecology and Power: Struggles Over Land and Material Resources in the Past, Present and Future*. Routledge, Abingdon, UK.

Robbins, P. (2004) *Political Ecology: a Critical Introduction. Critical Introductions to Geography.* Blackwell, Malden, MA.

Tsai, L.L. (2007) Solidary Groups, Informal Accountability, and Local Public Goods Provision in Rural China. *The American Political Science Review* 101, 355–372.

Wilhite, D.A. and Pulwarty, R.S. (2005) Drought and water crises: Lessons learned and the road ahead. In: Wilhite, D.A. (ed.) *Drought and Water Crises Science, Technology, and Management Issues*. CRC Press, Boca Rahon, FL, pp. 389–398.

Image courtesy of Tine Engedal

11 Social and Institutional Dynamics of Mobility as an Adaptation to Climate Change

Charles Aben, Esbern Friis-Hansen and Isaac Nakendo

Introduction

The recent surge in immigration to Europe from Africa and the Middle East has elevated cross-country migration to one of the top foreign-policy issues in Europe currently, as well as becoming a topic in the COP (Conference of the Parties) negotiations. This has greatly enhanced the focus of international debate on migration and its multiple causes, one of the most recently added causes of migration being climate change. The fear of an increase in what have been termed 'climate refugees' (Hampshire, 2002; Leighton, 2006) has been widely reported by international media and been used by international non-governmental organizations (NGOs) to argue forcefully for strong and immediate action to address climate change mitigation and adaptation (Christian Aid, 2007; IPCC, 2016). Estimates of the number of 'climate refugees' who will physically migrate as an effect of climate change before 2050 vary between 10 million (Jacobson, 1988) and 250 million (Myers, 1990), with some recent estimates going as high as a billion (Christian Aid, 2007).

Until now, no convincing evidence has been presented indicating that cross-border or even cross-continent climate migration is taking place on a large scale. Increasing local mobility as an approach to adapting to climate change is, however, a growing phenomenon that is largely overlooked in the international debate.

Understanding the relationship between climate change and mobility has evolved during the past two decades, from understanding climate change as a direct cause of migration to including multiple factors that are unrelated to climate change. For example, the Inter-governmental Panel on Climate Change (IPCC) adopted an alarmist position in the 1990s and predicted that people would be displaced by rising levels of seawater, droughts and floods. It later moderated its assessment and added social vulnerability and adaptive capacity to environmental hazards at the individual, household and community levels to its analysis (IPCC, 2001). Other possible explanations for migration include the availability of markets, access to infrastructure and the ability of the household community to cope with environmental conditions. Other factors are powerful drivers that reduce mobility, including improved access to social services such as health clinics, water supply and schools provided by government (Myers, 2002).

Meanwhile some literature, notably that on the 'environmental refugees' mentioned above, continues to argue that the environment is the direct cause of migration. The notion of climate refugees, however, underestimates the importance of historical migration patterns in high-risk environments, as well as the causes and processes of adaptation, and is largely at odds with empirical case studies showing the multi-causal character of migration patterns (Raleigh and

Jordan, 2010). The notion of 'environmental refugees' also wrongly portrays climate refugees as passive victims, rather than active citizens.

In recent years, scholars such as McLeman and Smit (2006) and Tacoli (2011a,b) have begun using climate change as an analytical lens through which to examine mobility and migration. Instead of viewing migration as a last resort, these scholars analyse mobility as a way in which local people diversify traditional agricultural livelihoods through adaptation to climate change hazards. In this study, we distinguish between: (i) everyday autonomous adaptation by people living in environmentally marginal areas; and (ii) responses to extreme climate change events such as extreme droughts.

By far most of the literature on drought and migration is concerned with the first point: everyday autonomous adaptation. There is a rich literature describing how rural populations, in particular pastoralists, living in environmentally marginal areas of Africa have developed a wide range of mechanisms over time to strengthen their ability to cope with slow climate change (Findley, 1994; Maxwell, 1996; Meze-Hausken, 2000). Communities such as pastoralists who live with environmental hazards as part of their daily lives engage in four sets of adaptive strategies, which they combine in different ways: (i) the cyclical migration of extended families together with their livestock in search of grazing niches; (ii) combining livestock rearing with crop cultivation through agro-pastoralism; (iii) diversifying livelihood activities by engaging in crafts or trade; and (iv) remittances (de Haan *et al.*, 2002). The last strategy, which involves long-distance labour migration, is critical because it provides a source of capital to invest in the three other strategies.

Mobility as an autonomous everyday adaptation strategy has multiple causes. For example Davies (1993) found that migration was one response in a basket of coping mechanisms in Mali. Enhancing physical mobility when faced with climate change hazards provides people with an opportunity to diversify their sources of income, thus spreading the risk, which in turn increases resilience for the household and/or wider family (Tacoli 2011a). Thus migration has become an increasingly important aspect of rural livelihood strategies in the face of the slow onset of climate change hazards such as desertification, soil degradation, the increased variation in rainfall patterns and temperature changes. In case studies in Bolivia, Senegal and Tanzania, alternative sources of income have increased migrants' ability to survive, while also increasing the resilience of their families back home through remittances (Tacoli, 2011b).

Much less has been written about the second point: the response to extreme climate change events such as extreme droughts. There is limited evidence from existing case studies to suggest that environmental hazards can cause changes to migration patterns (Raleigh and Jordan, 2010). Hunter (2005) reports that it is rare for the whole community to relocate physically in response to extreme drought. According to Ober (2014), however, although case studies of the recent past can serve as analogues of adaptation in the face of environmental crisis, they cannot accurately predict how climate change will influence landscapes and livelihoods in the future.

There is little doubt, however, that extreme climate change events could have such a devastating impact on people's livelihood security that traditional coping mechanisms, such as temporary seasonal migration, are no longer sufficient; migration involving the resettlement of entire households or communities might be required. Studies in Kenya and Somalia report that climate change in the form of prolonged droughts has resulted in dramatic increases in population around smaller market towns, reflecting increased dependence on food aid and available food in markets, as well as opportunities to work as casual labour (Little *et al.*, 2001).

During the early stages of a drought, pastoralists respond by following pre-established coping mechanisms, such as the long-distance migration of young men in search of employment that enables them to send their families remittances that contribute to their survival. If the drought continues, however, the remaining family, where women and children dominate, may engage in short-distance migration to the nearest location not affected by drought. Most such short-distance migration takes place internally, that is, within the area occupied by the same ethnic group or tribe (Henry *et al.*, 2004).

Study Focus

This chapter examines migration as an adaptation response to climate change hazards and analyses how climate-induced changes to settlement patterns influence livelihoods, as well as socio-economic and political processes and outcomes.

The overall objective is to explore the role of climate change on mobility in the cross-border areas of Teso and Karamoja that face frequent floods. Specifically, firstly, the study seeks to identify and document the causes of mobility in cross-border lowlands, the mobility patterns that can be identified and those who migrated. Secondly, the study sets out to document and analyse the consequences of mobility: how new forms of social organization have emerged, how resources are shared between immigrants and locals, the conflicts that have arisen and how have they been solved. Thirdly, the research seeks to determine to what extent mobility affects the environment and its effects on production and marketing systems. Finally, the research will assess the implications of mobility for service systems, culture and socio-economic benefits.

The enumeration area for the study comprises six sub-counties in the two districts of Katakwi and Napaka for the collection of field data. This was based on the fact that these were the major destinations and sources of migrants: Iriri, Lokopo and Lopeei sub-counties in Napak district represented Karamoja region, while Magoro and Ngariam sub-counties in Katakwi district represented the Teso lowlands. A field survey mission was conducted using semi-structured interviews with district and local officials from relevant departments, and three focus groups were held to discuss the above questions, which helped in gathering opinions from the affected areas.

Historical Context

The Iteso were originally part of the Karamojong who came together from Abyssinia in present-day Ethiopia several hundred years ago. The Iteso first settled in the present-day Usuk County. According to oral tradition, younger men and women always defied their elders in Karamoja and moved southwards during the dry season into present-day Teso before returning home during the wet season. They continued to repeat the process until they decided to settle there permanently. The Karamojong who remained referred to the migrants as *Atesin* or 'corpses' (the origin of the name *Iteso*) because they did not believe they would survive in the south, where they migrated.

Conflicts in this cross-border area are nothing new. On the eve of colonization, there were widespread regional wars, especially around the wetlands and no man's lands (Vincent, 1982). After colonization and independence in 1962, however, the Iteso and Karamojong lived harmoniously until the 1980s, when frequent droughts induced large-scale movements of the Karamojong into Teso in search of water and pasture for their animals. Cattle raids by the Karamojong against Iteso and surrounding areas are perhaps a major outcome of climate change in this part of Uganda. According to elders in Napak, cattle raids are nothing new either. The Karamojong clans even used to raid among themselves, especially after massive losses of cattle during long and severe droughts, at which times cattle-rustling was regarded as a traditional way of restocking. The demand for large numbers of cattle as dowry also seems to have fuelled the participation of young men in cattle-rustling.

According to residents of Magoro in Teso, the major cause of cattle raids was that whenever there was a drought in Karamoja the numbers of their cows generally decreased, and whenever the Karamojong reached Teso and saw cows actually resembling theirs, they admired them and took them by force. This created a continuous cycle of cattle raids because the Iteso had to find a way of recovering their cows. Cattle-rustling also became more violent, as the Karamojong acquired guns during the many security vacuums following military coups in Uganda, when the AK-17 replaced the traditional spear (Mkutu, 2006). This forced the Iteso from the border to migrate into government-protected camps. However, while most Karamojong have been disarmed over the past decade, the Iteson still consider the Karamoja region risky. Although the national camps for internally displaced persons have officially been closed, the displaced Iteso have only partially returned to their homesteads owing to the lack of confidence in the sustainability of the present peace.

With the Karamojong now disarmed, peace is returning to the region, and the Iteso, Karamojong and traders from outside the region are now crossing the local boundaries. Notwithstanding the fears over security, recent shifts in mobility patterns in the flood- and drought-prone cross-border lowlands represent significant changes in the area's socio-cultural and political landscapes. There have, however, also been marked changes in mobility patterns compared to the recent past, when droughts in Karamoja caused seasonal movements of people and their herds of cattle into the Teso lowlands. Now, the Karamojong are permanently occupying and cultivating large areas of Teso.

Communities have also moved during floods, when people living in low-lying areas have been forced to move and camp in the highlands. The displacement of people around the UPE Game Reserve to Angisa camp also represents a significant mobility on the part of communities. Mobility is not only a response to resource constraints; it can also worsen land constraints in the area to which the migration is taking place. However, it could also initiate institutional innovations in the areas affected. The socio-economic implications of these movements must be properly understood in order to design appropriate service delivery options for these cross-border lowlands.

Findings

The contribution of climate change hazards to enhanced mobility

Variations in seasonal weather patterns caused by extreme climate change events in the form of more frequent and more prolonged droughts have led pastoralists and their cattle to migrate further and further away in search of fodder and water. The major causes of mobility as reported by all respondents were extreme drought and floods. Karamoja often experiences extremely dry conditions. The most remembered droughts were in 1999 and 2007, when people lost their animals, crops and lives. The District Production Officer of Napak, Mr Lodungokol John, reported that the weather pattern in Karamoja has changed, with unreliable rainfall affecting production. As a result, people are forced to move to other areas to obtain pasture for their animals or grow crops. This is the main reason why Iriri sub-county receives large numbers of migrants. Cattle-herders have been forced to go and settle along the borders of Teso to ensure they can always easily enter Teso for pasture (Fig. 11.1). For example, Karamojong from Amudat, Nakapiripiriti and Napak camp in Angisa in Magoro. The most interesting issue in the climate refugee situation is that the Karamojong sometimes claim they are following their own water flowing into Teso region in the form of floods. After the rainy season, which is accompanied by flooding, the Karamoja region suffers drought. The Karamojong use this as an excuse to follow their water that has flowed into Teso in the form of floods. This has therefore become a source of conflict because, whenever the Karamojong claim their water, the people of Teso are compelled to remember their cows, and also their children not going to school.

The study also found that, after the government curbed cattle-rustling, the reasons for the Karamojong moving changed because they now move into Teso and the green belt of Napak district to look for paid work. This is a major change from the past. One respondent explained that in the past the Karamojong have always moved to find water and pasture for their animals. This was very intense in the 1980s, when there was no order in the country and cattle-rustling was at its peak. However, there have been times when the Karamojong have been restricted in their own region with their cows and the government has used the police and army to monitor the boundaries continually. This caused a great need for alternative livelihoods and has forced them to move to Teso to work in exchange for food and money. Reportedly, it is mainly the women, however, who move in order to look for food, the men staying at home as cattle-herders.

The study also found that, as a result of climate change and the restriction of cattle-rustling by the Karamojong, for most people the number of cows they have has declined tremendously. There is still some livestock stealing, although on a small scale. People in Angisa in Magoro sub-county reported that some thieves from Karamoja can move as far as from Nakapiripirit to Katakwi to steal cows and goats, taking advantage of the bush that has developed in the Angisa

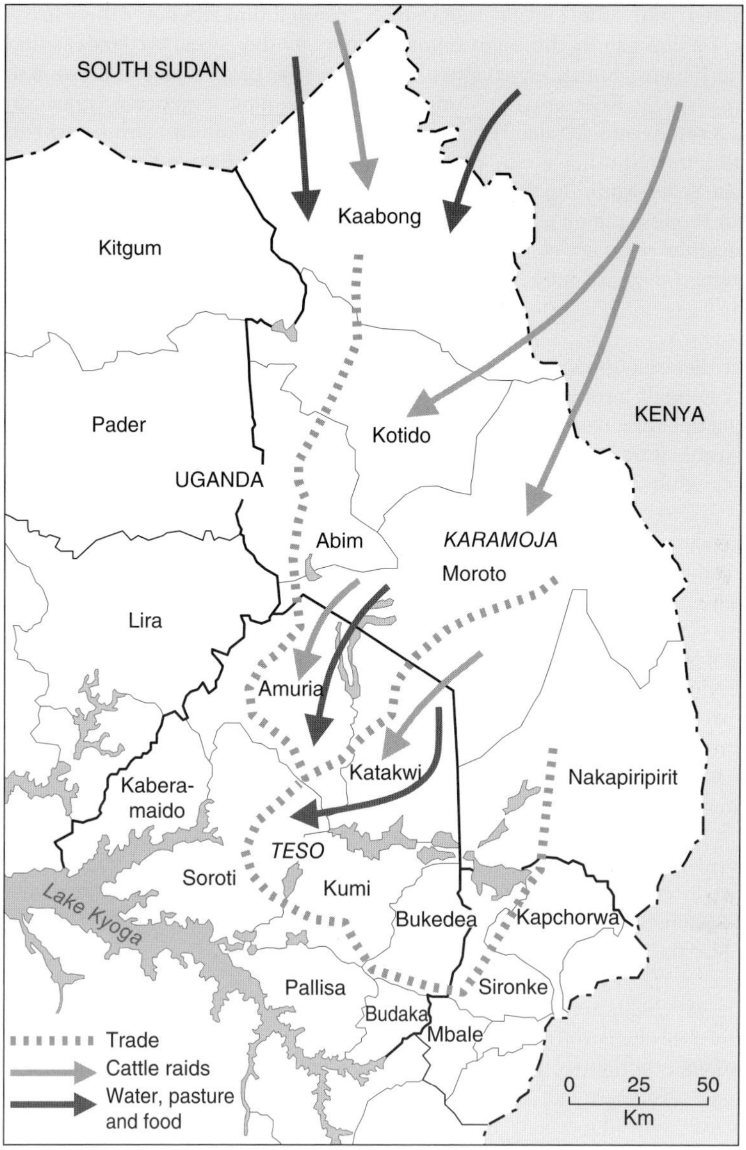

Fig. 11.1. Climate change-induced mobility patterns in Karamoja and Teso regions, Uganda.

area, which the government has demarcated as a game reserve, to move the animals. It was also noted that people in Teso, rather than just accepting this, are involved in stealing cattle from the Karamojong. However, by the time of this study, the government had placed police and army to control this small-scale animal stealing that was turning into a civil challenge for local governments in the area. From these findings, therefore, it can be noted that, among the Karamojong tribes, the main causes of mobility have been the need to obtain water and pasture for the animals, the need to restock after the herds have been reduced by climate change and the need to acquire for food from the neighbouring communities.

In Teso, the research uncovered a number of climate-related reasons for mobility, the most obvious being the floods that affect the area almost every year. The climate in Teso is often associated with the Awoja wetlands, which cover 14 districts in the area, namely Soroti, Amuria, Katakwi, Napak, Kween, Bukedea, Kumi, Serere, Ngora, Kaberamaido, Amuria, Kapchorwa, Sironko and Mbale. This area is low-lying compared to other areas, especially the Karamoja/Sebei and Kapchrwa regions. This has made the Teso region prone to flooding from these neighbouring areas, and floods are the most common cause of mobility in the following ways:

1. Whenever the area floods, people are forced to move their animals from grazing land in the swamps to the uplands. This is a very big challenge for migrant cattle herders from other districts because, while the swamp is communal grazing land, different clan members own the upland. This means they are moving into areas they do not own. This creates further movement until they reach areas where they can graze without being disturbed, which is not easy because in Uganda today there is no land that someone does not claim.

2. Often, when there are floods, people move from their homes to internally displaced persons' camps, which are supported by the Ministry of Disaster Preparedness and other donor agencies. This produces further movements because some organizations select certain children to be allowed to continue their education in other places in sometimes far-off districts. Some people from Teso have been allowed to join traditional displaced persons' camps such as Kiryandongo, which was traditionally a camp for refugees from armed rebellions in Uganda.

3. The two mobility situations in Teso just mentioned are more permanent but certain temporary movements were discovered during the study. For example, it was found that in Angisa during the floods, different families are affected differently. There are some whose houses are almost submerged, and others whose houses are raised a little, where the flooding remains half under the beds. Those whose houses are prone to submerging move to their neighbours whose houses are raised and remain there until the end of the flooding season. This was a temporary movement but with appalling social and economic effects on the population.

4. The other mobility pattern involved fishermen dependent on fishing as a livelihood. It was reported that whenever there are floods a large part of the Awoja wetlands is flooded. During this time Lakes Opeta and Bisina join up to become a new, larger lake created by the floods. This means that the fish become disturbed and move from place to place looking for safety and cannot easily be located by the fishermen. Hence, the latter must look for alternative sources of income until the end of the floods. The fishermen therefore move to other places for either agriculture or trade purposes until the end of the flooding season. This is a serious mobility pattern because it involves entire families of fishermen.

5. It was also found during the study that, even within the Teso region, there were serious movements revolving around social services. Reports indicate that, during the floods, most health units are either destroyed or their access becomes blocked by the floods. It was also reported that every family had secured a small boat to help them access health services elsewhere during the floods. This problem affected not only health services, but also markets, schools and entertainment facilities. It was reported from Angisa parish in Katakwi that, due to the frequent flooding of the Awoja wetlands that leads to the parish, there is no restaurant, church or entertainment, and that people in search of these services must move elsewhere in Magoro, or even as far away as the Amuria district townships.

Changing mobility patterns as adaptation to climate change hazards

Table 11.1 shows the reasons for changes to climate mobility, and a number of new reasons have emerged. It should be noted that, concerning the Karamojong in the past, their main reason for movement was the search for water and pastures. This is because cattle-keeping was their main source of livelihood. They were used to drinking the milk and blood of their cows, and sometimes exchanging milk for other food supplies on a small scale. They never thought of staying in one place and cultivating their own food. The cow was therefore considered the Karamojong's lifeline, and whenever they moved from place to place they would steal

Table 11.1. Changing mobility patterns in Karamoja and Teso regions, Uganda.

Mobility patterns	Traditional reasons for movement	New reasons for movement
Karamojong moving to Teso region	Search for pastures and water and restocking after number of animals reduced due to climate change	Search for labour, food, social services and restocking
Teso moving to Karamoja	Recovering animals stolen by Karamojong, small-scale trade	To recover animals stolen by Karamojong or intermarriages between the two tribes, trade, missionary work
Teso people moving from lowlands to uplands	Searching for flood-free places, health services, education services, grazing, cultivation and markets	Searching for flood-free places, health services, education services, grazing, cultivation and markets
Pastoralists from Nakasongola, Luwero, Kumi and other neighbouring places	Never used to come to Teso	Searching for pastures and water

other people's cows in order to accumulate as many cows as possible. This made them live a life defined by the nature of climate change, and wherever the climate took them, there they went. They were also known as warriors because, wherever they went, they were tempted to steal other people's cows. This sometimes involved serious fighting with other tribes, but the Karamojong were usually much more organized because they would plan their attacks properly and were well equipped, being involved in trading in firearms with the Sudanese and Kenyans. During this period there were many rebel activities, and many people survived by selling guns and bullets to the Karamojong. One respondent revealed that bullets were sold in the way other products such as beans were sold in Karamojong markets. This made the Karamojong confident about the legality of their trade and they never worried about attacking a community to stock up on cows. This meant that if one wanted to know where a Karamojong was, one would just feel the temperature. The higher the temperature, the further the distance of the Karamojong from Karamoja.

Although it is true that the Karamojong still look for water and pastures, their movements have now changed because they are restricted to their own regions and only come to the boundaries of other tribes to trade. This is because government interventions control their movements in a number of ways:

1. The Karamojong have been introduced to growing crops, especially climate-resistant ones such as cassava.

2. Schools have been introduced in Karamoja, which has controlled Karamajong movements because government has made universal primary education (UPE) mandatory to the extent that a parent who fails to comply is penalized. It is hardly surprising that the Karamojong are the people with the lowest number to have reached university level in Uganda.

3. The government has also introduced improved breeds of cows to the Karamojong, though the latter have stuck to their local breeds. None the less, this has helped control their movements.

4. The government has also built infrastructure, especially roads leading to urban centres and markets, thus introducing alternative livelihoods to the Karamojong.

Despite these initiatives, however, the Karamojong are still mobile, but for different reasons. For example, in Iriri sub-county the pattern of people coming from other sub-counties and districts of Karamoja increases from year to year. This is because people are moving their households to cultivate crops and to obtain pasture and water for their animals. While the pattern of movement from Karamoja to Teso is gradually changing compared to the traditional seasonal movements, cattle herders still move closer to the border of Teso during the dry season.

The number of people, especially women and children, moving from Karamoja to Teso and other regions is increasing and no longer depends on the season because they are looking for casual labour at most times of the year. The most significant aspect of these movements was found to be that in the past other tribes were as hostile to Karamoja as the Karamojong were hostile to them. Things have changed, however, because now the Karamojong intermarry with other tribes. A respondent in Wera sub-county in Amuria district revealed that when Karamojong women come for work, some of them are married to Teso people. Karamojong boys also go with Teso girls, and dowries are exchanged between the two tribes. This marks a very big change in the history of climate change mobility between the two tribes and is a positive trend. It was also revealed in the study that there are many Karamojong children studying in Teso schools. Some of these children are working casually in Teso and paying secondary school fees. This is important because schools in Karamoja are still very few compared to other areas. This is also a reflection of the new patterns of mobility that are emerging as adaptations to climate change.

One other new mobility pattern that has emerged is the movement of people from other regions to the Karamoja region. In the past, the movement was practically unidirectional, involving only movement from Karamojong to other regions. However, today people are increasingly moving in response to climate change hazards. The mobility of people from other parts to Karamoja has particularly been seasonal, especially during the dry season, when there is hunger and scarcity in Karamoja. This is the time when traders from outside Karamoja sell produce such as beans, sorghum, cassava and other goods required by the Karamojong. This has now intensified, however, because of the interventions that have taken place in Karamoja such as the development of transport networks, urban development and the rise of industries involving different types of mining and processing. Although all this is happening, Karamoja still suffers from the lack of essential foodstuffs owing to the nature of its climate, thus paving the way for movements of traders into the area, creating yet more new mobility patterns. This has further inspired missionaries and philanthropic organizations to undertake a sort of Great Trek into Karamoja. Karamoja generally has a reputation as somewhere hostile and feared. Not even missionaries ever managed to bring the Karamojong together and spread the gospel among them. However, thanks to a number of innovations, there are now gathering places and missionaries, so NGOs and other organizations are using the opportunity to work with the Karamojong on climate change adaptation. This is an important mobility pattern that may lead to many other new changes.

It should not go unrecorded that mobility patterns have also contributed greatly to changing the Teso region and actually led to further mobility into the area. Enhanced traditional patterns of seasonal migration are creating demographic changes, including increased urbanization, mostly in the development of small towns along the traditional boundaries. This especially concerns the growth of towns around the markets that have been created by people coming together as a result of climate change. These areas include the Katakwi trading centres, especially in Magoro, the Wera trading centre in Amuria and others. These trading centres continue to attract people from neighbouring areas to exchange commodities. As a result of climate change, different communities have different foodstuffs, and this gives them an opportunity to exchange what they have for what they do not have. It is not only this direction of transformation, but local government roles in harmonizing land tenure that have been put to the test. As a result of climate change over a wide area, members of local communities have been tempted to sell land to move to other areas where the situation is better. Because of the need to protect land, however, clan leaders have strengthened the rules concerning the selling of land. Thus one cannot sell land to a non-clan member as a priority, as occurs in Soroti, Amuria and Katakwi. It was revealed that before any land is sold the procedure should be as follows:

> The seller makes an application to the clan leaders, who meet to consider the reason for the sale, and the buyer is also assessed. He should be a clan member, though a member of another clan is allowed if no clan member wants to buy the land. However, he must bring a letter from his clan to show he is a good person. If there is no member available from another clan, the land will not be sold.

These regulations have affected individuals who want to sell their land and move to areas where the climate challenges are still minimal. It has also affected the government's ability to develop land either by introducing new investors or the government itself taking initiatives. A good example is Angisa parish, which Uganda Wildlife Authority declared a game reserve and where the game reserve was demarcated, but people have refused to leave the land. The clan leaders have insisted that the land of their ancestors cannot be taken by the government. At the time of this study, Uganda Wildlife Authority had demarcated the reserve and pushed the people to land near the Awoja wetland in Angisa parish but the clan leaders were still resisting leaving their traditional land.

The demarcation of borders by government to create clear districts has revealed another pattern of border mobility that is well defined. Many tribes now find it very hard just to move into another people's district and graze there. As a result, there is a growing movement of cattle herders towards the borders of other districts in search of water and pastures. These migrants enter the district but remain on the border because they know there is a likelihood of conflict. Within Karamoja, for example, people move towards the green belt, where the climate is fairly good. Iriri sub-county in Napak and Abim Districts are the major destinations. Some people, especially the cattle herders, have settled along the borders of Karamoja and Teso, where they easily cross into Teso looking for pasture and water. These areas include Lokopo sub-county and Lorengecora. Other people have moved into Teso, affecting the sub-counties of Ngariam, Magoro, Usuk and Agogoja in both Katakwi and Amuria Districts. Another group has moved into towns like Soroti, Mbale, Iganga and Kampala looking for casual work. In Soroti they live in the area called Kichinjaji, while in Iganga they live in the area called Kikoromojo. Some of these people, as noted earlier, come as far as Nakasongola and Luwero Districts looking for water and pastures in the Awoja wetlands. Though the wetlands are usually affected by flooding, they have the capacity to provide pastures and water throughout the dry season. This has attracted different groups who have decided to remain in the boundary of Teso District grazing in the wetlands. It was revealed that, because there is no grazing area in the uplands (because the population has left the lowlands for the upland), everybody now has shifted to the wetlands for grazing. The wetland therefore now has a mixture of animals, including cows, goats, sheep and even pigs. Grazing in the swamp is free range, with some people just leaving cows in the swamp because sometimes they come from long distances with their cattle, go back home, and come back after a week to check on them. This form of mobility and adaptation to climate change has disadvantages because there is increased theft of animals and animal diseases, and it encourages irresponsibility among farmers.

Resource conflict and reconciliation

In the study, it was found that all the above mobility and adaptation practices have led to conflicts over resources and processes of reconciliation. In the first place, as a result of drought in, for example, the Teso region, people moved into to the wetlands mainly to graze their animals. The fact that everybody from far and wide looks at the swamp as the only option in areas where there is no demarcation, as Omugenya in Gweri sub-county in Soroti and parts of Amuria, there is a serious conflict between cattle herders and cultivators. It was revealed that in some places, cattle herders deliberately come at night to enable their cattle to graze in other people's gardens within the swamp. The intention is to stop them from using the swamp for cultivation because they feel they need it for water and pastures. It was also revealed that the ownership of the swamp is such that those who have gardens near the swamp are the owners of the swamp and had also put restrictions on entrance by cattle herders. Cattle herders, however, believe that the wetlands belong to the government and there should be equal use rights to them. These different perceptions have been the main source of conflict between cultivators and cattle herders.

The other source of conflict in resource use is between the traditional leaders and clan members, especially young people. It was found out that, as a result of climate change, some members of society, especially the young, want their share of land to sell and move to other areas. However, the traditional leaders' parents are refusing to give land to the children. This has become a serious challenge that in some homes has led to domestic violence, including killings of parents by their children. In solving these challenges, however, the traditional leaders eased the process and allowed children to sell land but to do so to clan members. NGOs also provided the children with counselling and encouraged them to work hard patiently to see if things changed, rather than cause trouble in the family.

The other significant conflict over resources that was observed was between women and men, especially when it came to inheritance, where women were not acquiring land through inheritance. They were expected instead to obtain land from their husbands, as had been practised for some time. However, as a result of climate change, some married women have returned to working the land of their parents. Sons reject this and argue that women should not be given land. So far there is no agreement that a woman can be given land but instead she can only till it; she cannot own it, because it remains in the possession of the family while she is not using it. This is an important climate resource challenge that needs to be handled carefully by policy makers in the interests of the effective inclusion of women.

Inter-tribal resource conflict occurs when members of certain tribes cross into other tribes' territories to steal animals. This was said to be a form of cattle-rustling but on a small scale. It was reported among the people of Katakwi, especially in Angisa, and among the Karamojong, as well as among the people of Soroti and the tribes coming from outside to graze in the Awoja wetlands. In order to solve this problem, a local ad hoc committee was created, which adopted just one rule about stealing animals: 'steal one, if caught you pay four'. This rule is not negotiable to the extent that if a cow of a certain community is found in the kraal of another community – a kraal is owned by eight people or more – there will be no discussion about who stole it and any four cows will be selected from the kraal as compensation. This created more conflicts, however, because the members of the kraal would start fighting amongst themselves over who brought the stolen cow to the kraal, forming the basis for suspicions and conflicts among friends and relatives within the kraal. The conflict also remains between the community that has been compensated and the compensating community to the extent that the compensating community keeps praying for bad luck to happen to the compensated community so they can retaliate.

Resource conflicts also existed between the government and communities themselves. There are many instances where the government had demarcated land, but the people had continued to encroach on it, as happened in Angisa parish in Katakwi. Encroachment on to many wetlands was also reported, causing conflict between the community and local leaders.

It should be noted that, in the management of mobility and climate change, national politicians and local politicians have responded differently. For example, the people of Amuria revealed that the issue of wetland demarcation was not necessarily an environmental concern but was due to conflict between cattle-herders and cultivators. Those seeking election, especially members of parliament, were at the forefront of solving conflicts over demarcation for the sake of their own political advantage. As a result of this approach, wetlands where there are no such conflicts have not been demarcated, raising the question of whether the politicians care about long-term environmental conservation or short-term human survival. If this approach is evaluated properly, short-term human survival is the answer.

As a result of the need for resources in the context of mobility and adaptation to climate change, local politicians do not prioritize the need to implement government programmes but think they should protect the people from hunger first. This has created a problem between local leaders and their relatives and friends because the former also use these resources themselves and therefore find it hard to intervene when the rules are violated. Certainly the environmental committees created at the parish level are doing a significant job in sensitizing the people, but are constrained because they are not paid and were given just a bicycle to monitor the environment in the entire parish, which is not easy. From the above findings, resource conflict and reconciliation are processes that are acted out in parallel with each other.

In discussing the issues of mobility and climate change, the context of the recent civil war is important. Uganda has been going through a civil war from the 1980s until now because the rebels who have been terrorizing the country are still in existence, though the guns are now quiet. Some of the mobility has been precipitated by the civil war, to the extent that one might ask whether climate or the civil war has had the greatest impact on mobility. Given that people have suffered in the war as well as as a result of climate change, they now prefer co-existence to violent conflict. This is the reason different tribes can graze in one and the same wetland without major tribal conflicts. It is also the reason politicians and environmental management authorities are simply looking on while the environment is being degraded. The governance system has failed to implement environmental management policies because the current imperative is for co-existence rather than conflict.

Socio-economic synergy from ethnic co-habitation

In this chapter, we should highlight the complementarity of the livelihoods of the people of Teso and Karamoja that, in one way or another, have eased climate change adaptation through mobility. It should be remembered that, before the National Resistance Army war that brought President Museveni to power started in 1981, the people of Teso were the biggest cattle herders in Uganda. In fact, unlike the Karamojong, the farmers in Teso were a little more organized and focused. They did not move as far while grazing as the Karamojong and Banyankole and gained a lot from cattle. However, the major part of the war was fought in Teso, where farmers were killed and their cows taken by the fighting forces. Even in 1990 the government promised to restock the Teso region but nothing has happened yet. This means that the Teso and the Karamojong had something. One elderly man from Katakwi reported that 'the Karamojong people raided cows, and the Teso people raided Karamojong cows in return. However, the Teso cows were finished by the rebels, not Karamojong, and the Karamojong cows were finished by the Kenyan Turkana, not the Iteso. This is the reason the Karamojong and Teso people co-exist these days'.

Co-existence has been made possible by the fact that the Karamojong have been predominantly cattle herders, whereas the Teso have mixed agriculture. For example, the Teso are the biggest growers of cassava, groundnuts, simsim, graingram, sorghum, beans, maize and other crops that do not do as well in Karamoja. This has created a competitive advantage for the two communities, where the Karamojong can supply animal products such as milk, ghee and skins and hides, whereas the Teso supply them with crops. Karamojong cows are of great value, especially for ox-ploughs, because they are usually hired by the people of Teso to cultivate their land, sometimes for free, because the Karamojong do not see the value of cultivation as much as the Iteso. This has enhanced co-existence between the two communities.

It should also be noted that the people of Teso have benefited from the manure generated by the Karamojong cattle, which is used in their gardens and which they do not need to buy because the Karamojong do not need it. This has strengthened the need of the people of Teso to stay with the Karamojong. This analysis indicates that the Teso people and the Karamojong need each other and that this has eased climate mobility in the circumstances of adaptation to climate change. Finally, as noted earlier, there are limited social services in Karamoja in the form of schools, hospitals and entertainment. The Karamojong cannot remain in conflict with the people of Teso, where these services are comparatively well developed. These have become important socio-economic synergies creating strong ethnic co-existence in the context of mobility and smooth adaptation to the challenges of climate change.

Institutional change and climate-induced mobility

Associated with mobility has been the emergence of new social institutions, community-based organizations (CBOs) and so on. The increased need for social services has led to shifts in focus and the emergence of new projects among NGOs such as TPO and LWF. Market institutions have also emerged to handle the increased movement of commodities in this cross-border area. It was found that in most sub-counties, local council minutes show increased deliberations over how to maintain peaceful co-existence between the Iteso and Karamojong, and how to reduce inter-tribal conflicts. In Magoro sub-county, a team has been formed to handle conflicts with the wildlife authorities and decide how to reinstate the 12 villages of Angisa parish that have now been reduced to only two villages. In the border districts of Napak and Katakwi, there is increased monitoring of livestock diseases by the veterinary officers of the districts, who regard the joint protection of the herds of the two tribes as a strategic priority. In the Gweri swamps in Soroti, an ad hoc committee has been set up to solve problems arising from the joint herding of animals in the wetlands.

Political implications of mobility

Mobility in cross-border areas has been accompanied by conflicts over land and other natural resources. These conflicts have been politicized in the decision-making arena at the local government level. Politicians have to tread carefully when handling conflicts, as they can lose their leadership positions at election time because of a failure to act decisively to solve resource conflicts between settlers and the native population. However, even politicians who contribute positively to reaching compromises between different ethnic or social groups can also lose leadership posts during elections because they might be seen as traitors.

Discussion

Mobility has resource implications. Resource sharing between immigrants and local people depends on ethnic perceptions. Resource sharing among the Karamojong who move internally from dry areas to the green belts on the border with Teso was reported as not having created any challenge simply because the major resource, namely land, is communally owned. It was reported that migrants are looked on as brothers and sisters. As long as they comply with the community's principles they are therefore easily integrated. Some migrants do not stay long, coming and going depending on seasonal changes. In Teso it is a different story because there are conflicts between the indigenous people and the migrants. The Magoro sub-county chief reported that migrants from Karamoja forcefully occupy land, destroy local people's food and steal properties using spears and guns. However, these conflicts have been managed through negotiation with the leaders of both Karamoja and Teso, with areas like Angisa being left to them as a solution. Those who have gone to the towns are living in abandoned government-owned areas, where they base themselves to look for casual work for their survival. As in Kampala, the government has adopted a policy of picking children from streets and taking them back home or to rehabilitation centres, and they are working with leaders in Karamoja to stop their movement to towns by engaging them in agriculture and similar occupations.

The conflicts that arise are solved in different ways. In Karamoja no conflict was reported as arising out of the presence of the settlers and if any does arise it is handled communally. In Teso, the conflict that emerged first was the border conflict in Napak and Katakwi Districts. A claim has been made that Karamojong have occupied land in search of water and pasture at some point beyond the traditional boundaries separating Teso and Karamoja. An independent firm was hired by the government to demarcate the boundaries again, but there are still disagreements, with many political implications for leaders who are manipulating the situation for their own political interests. The second conflict was between Karamojong cattle herders and the wildlife authority, where Angisa, the famous parish grabbed by migrants in Magoro sub-county, was gazetted land owned by the Uganda Wildlife Authority (UWA). As a result, the UWA claimed the land and forced the migrants to leave. This has worsened the relationship between Teso and Karamoja, however, because during the dry season cattle herders come and destroy the gardens of the Iteso in retaliation, and this keeps the security forces busy in the area, as the local sub-county chief reported.

Mobility also has effects on the environment. Mobility has negatively affected the environment in both Karamoja and Teso and even beyond. The environmental officers of both districts reported that wetlands are being destroyed in search of water and pasture, as well as in converting land for agriculture. They gave the example of the Apetelolim wetland, where migrants cut trees for charcoal burning to earn a living. Migrants have encroached on gazetted government land in Angisa, Katakwi and many other wetlands.

Another consequence of mobility is its socio-economic impacts. In Iriri-Napak District, mobility has helped increase production. Farmers have adapted new farming methods and improved crop varieties of maize, cassava, sorghum and beans. Produce prices have fallen somewhat compared to before. For example, the sub-county chief reported that a kilo of *posho* that used to cost UGX4500 or UGX5000 (Ugandan shillings) is now sold at UGX3000 during the harvesting period. Farmers have also adopted improved breeds of animals, mainly goats and cattle. It was reported that at the border markets, Karamojong farmers are exchanging local cattle for improved heifers and goats from Teso. Market structures have been established, and now the District Production Officer of Napak District sees the sub-county as the food basket for Karamoja. In Teso, however, the contrary is the case. Production levels in the affected area have drastically declined owing to insecurity and tensions caused by migrating Karamojong. The animals of migrants destroy gardens, and investors are scared away, while commodity prices have increased. Granaries, the traditional storage systems, are being destroyed and the Karamojong steal food. According to the sub-county chief, this has affected development because local people have abandoned the place and moved to neighbouring districts.

The implications of mobility for social services are also noticeable. In Iriri sub-county, health and education services have been overwhelmed by the rising level of population. Medical staff and teachers cannot satisfy the demand for services, resulting in increased illness and death rates, as reported by the Community Development Officer (CDO) of Iriri sub-county. This has also increased school dropout rates and poor performance in schools. In Teso areas mobility poses challenges for public servants. Government officials posted to these areas fear working there because of insecurity. Field staff spend less time in the field owing to fears of an attack, reducing their performance levels. Both Magoro and Ngariam sub-counties reported high turnover rates for public servants. There is low class attendance in schools, which both sub-county chiefs and the CDO attributed to fears of Karamojong migrants.

Finally, mobility has implications for the cultural systems of immigrants and hosts alike. District staff and those at the sub-county level in Karamoja appreciate mobility because it triggers good changes in culture. The disarmament process has made raiding impossible, the ownership of guns is illegal, and peace and security have finally been brought to the region. Mobility has also improved the dress style of the Karamojong and formal education has improved.

Conclusion

Climate change has had far-reaching effects on the flow of both resources and people across the Teso–Karamoja boundary. Although there are some negative consequences, mobility has helped the two communities forge new relationships that will help in the long-term adaptation to droughts that have been the main cause of conflict in this cross-border area. The institutional changes arising from new socio-economic and socio-cultural forms of interaction need to be harnessed to aid long-term adaptation to climate change.

References

Christian Aid (2007) *Human Tide: The Real Migration Crisis. A Christian Aid Report*. Christian Aid, London.

Davies, S. (1993) Versatile livelihoods: strategic adaptation to food security in the Malian Sahel. Report to ESCOR Overseas Development Administration, LOS, Brighton, UK.

de Haan, A., Brock, K. and Coulibaly, N. (2002) Migration, livelihoods and institutions: contrasting patterns of migration in Mali. *Journal of Development Studies* 38, 37–58.

Findley, S. (1994) Does drought increase migration? A study of migration from rural Mali during the 1983–1985 droughts. *International Migration Review* 28, 539–553.

Hampshire, K. (2002) Fulani on the move: seasonal economic migration in the Sahel as a social process. In: de Haan, A. and Rogaly, B. (eds) *Labour Mobility and Rural Society.* Frank Cass Publishers, Cornwall, UK.

Henry, S., Schoumaker, B. and Beauchemin, C. (2004) The impact of rainfall on the first out-migration: a multi-level event-history analysis in Burkina Faso. *Population and Environment* 25, 423–460.

Hunter, L.M. (2005) Migration and environmental hazards. *Population and Environment* 26, 273–302.

IPCC (Intergovernmental Panel on Climate Change) (2001) *Climate Change 2001: Synthesis Report*. Cambridge University Press, Cambridge, UK.

IPCC (2016) *Fifth Assessment Report (AR5)*. IPCC, Geneva, Switzerland.

Jacobson, J. (1988) Environmental refugees: A yardstick of habitability. *Worldwatch paper 86*. Worldwatch Institute, Washington DC.

Leighton, M. (2006) Desertification and migration. In: Johnson, P.M., Mayrand, K. and Paquin, M. *Governing Global Desertification*. Ashgate Press, London, UK.

Little, P.D., Smith, K., Cellarius, B.A., Coppock, D.L. and Barrett, C. (2001) Avoiding disaster: diversification and risk management among East African herders. *Development and Change* 32, 401–433.

Maxwell, S. (1996) Food security: a post-modern perspective. *Food Policy* 21, 155–170.

McLeman, C. and Smit, B. (2006) *Migration as an Adaptation to Climate Change*. Department of Geography, University of Guelph, ON, Canada.

Meze-Hausken, E. (2000) Migration caused by climate change: how vulnerable are people in dryland areas? *Mitigation and Adaptation Strategies for Global Change* 5, 379–406.

Mkutu, K.A. (2006) Small arms and light weapons among pastoral groups in the Kenya-Uganda Border Area. *African Affairs* 106, 47–70.

Myers, G.M. (1990) Optimality, free mobility and regional authority in a federation. *Journal of Public Economics* 43, 107–121.

Myers, N. (2002) Environmental refugees: a growing phenomenon of the 21st century. *Philosophical Transactions of the Royal Society: Biological Sciences* 357, 609–613.

Ober, K. (2014) *Migration as Adaptation: Exploring Mobility as a Coping Strategy for Climate Change*. The UK Climate Change and Migration Coalition, Oxford, UK.

Raleigh, C. and Jordan, L. (2010) Climate change and migration: emerging patterns in the developing world. In: Mearns, R. and Norton, A. (2010) *Social Dimensions of Climate Change Equity and Vulnerability in a Warming World*. World Bank Publications.

Tacoli, C. (2011a) Migration and mobility in a changing climate: a policy perspective. *Revista Interdisciplinar da Mobilidade Humana* 36, 113–124.

Tacoli, C. (2011b) Not only climate change: mobility, vulnerability and socio-economic transformations in environmentally fragile areas of Bolivia, Senegal and Tanzania. International Institute for Environment and Development, London. Available at: http://pubs.iied.org/pdfs/10590IIED.pdf (accessed 18 January 2017).

Vincent, R.G. (1982) *A Handbook of Environmental Economics, Vol. 3*. Elsevier, The Netherlands.

Image courtesy of Tine Engedal

12 Hoarders as Saviours: The Performance of Regional Grain Traders During and After Extreme Drought

Godfrey Suubi and Esbern Friis-Hansen

Introduction

Over the past decade, Uganda has experienced more severe and frequent droughts, a development which has been attributed to climate change (USAID, 2013). As with farmers all over the world, those in Uganda are highly dependent on rainfall. This is also the case for Soroti District, in the Teso sub-region, the case study for this chapter and a district that experienced a failure of rainfall during the second half of 2013. These extreme climate change events represent a significant threat to the livelihoods of the smallholders who produce most of Uganda's agricultural production. Agricultural production is prone to drought, which affects both producers and consumers; the former loose out when prices are low, the latter when prices are high.

The responses to drought from central and local government have been of limited scope and mostly technical in nature. The focus on the centrally governed National Agricultural Advisory Services (NAADS) has been on delivering seed to farmers who have lost their entire crop to drought. Longer-term indirect support includes the National Agricultural Research Organization (NARO), which is engaged in breeding short-season and drought-tolerant varieties, and Uganda's meteorological institute, which offers an early warning system. Only a minority of farmers have access to such assistance, however, and its effectiveness as an adaptive measure to extreme drought is very limited.

The focus of this study is the role played by grain marketing during situations of drought. Since 1987, Uganda has pursued economic reform programmes that have resulted in the adoption of more liberal agricultural trade policies. The aim behind the adoption of liberalization was to reverse the state monopoly in internal and external agricultural trade and price controls pursued by previous governments (Friis-Hansen, 2000). The assumption was that Uganda's agriculture had stagnated because the state had discouraged farmers from producing because of the government's monopoly of trade in agricultural products through its parastatal marketing boards. At the time these boards were perceived to be corrupt and inefficient, resulting in farmers not getting paid on time. Uganda's open trade policy for food staples has enabled its traders to offer products and services competitively, reliably and sustainably, and Uganda is now a food basket for East Africa. There are no export restrictions on agricultural products nor has the government instituted any recent ban on trade in food. Consequently, the flow of maize from Uganda to Kenya is one of the larger and more consistent cross-border trade flows in the region, amounting to approximately 120,000 tons per year. There is also cross-border trade with Rwanda at 50,000 tons and South Sudan is becoming a growth market for Ugandan products.

In effect, liberalization policy was meant to allow produce prices to be determined by market mechanisms that would incentivize famers to improve agricultural production and with it increased incomes and better food security. The World Bank Uganda Diagnostic Trade Integration Study (World Bank, 2006) and the Ministry of Tourism, Trade and Industry (Ministry of Tourism, Trade and Industry, 2007) both recognize that trade policies can have important implications, both positive and negative, for smallholders in fighting poverty. In theory, market liberalization and effective competition in the trade in agricultural grains should level out the effects of a supply shock following a drought. Firstly, farm-gate prices for food products should rise as supplies are reduced, reducing the hazards associated with drought and low harvests for farmers, including the poorest and most vulnerable. Secondly, at the wholesale level, price increases should be dampened by competition, leading to inflow of grains from regions with surplus production.

Study Focus and Methodology

Although market liberalization of the trade in agricultural gains in theory contributes positively to climate change adaptation by lowering the hazards associated with drought, empirical observations in Soroti District during the 2013 drought suggested that the absence of trade regulations enhances the hazards associated with drought. This chapter examines the relationship between trade and climate change and shows how regulating trade can be a tool for climate change adaptation. The purpose of this study is to examine food commodity hoarding in circumstances of extreme climatic change and address the role of local and meso-level institutions. The chapter will focus on the extent to which climate change hazards associated with drought are influenced (enhanced) by the behaviour of private-sector traders. Using the 2013 drought in the Teso region of Uganda as a case study and fieldwork conducted in Soroti, the chapter will examine:

1. How the drought affected the vulnerability of the rural population (poor harvest, pressure to sell early to meet the costs of social reproduction).
2. The extent to which aggressive early-buying traders take advantage of farmers' vulnerability, increasing household food insecurity.
3. The extent to which traders become hoarders, as they delay selling their stored grain in a situation of deficit food supply (regional market failure).
4. How traders use early warning data to optimize profit by acquiring loans to purchase aggressively pre-harvest and to optimize selling when prices peak.

This chapter uses secondary literature on liberalization and agriculture/food security located in libraries and on the Internet. Statistics were collected from the Bureau of Statistics (Entebbe), the Agricultural Secretariat (Bank of Uganda) and Departments of Agriculture in Soroti district. Secondary data were mainly concerned with the historical, including recent trends in production and the market trends of specific enterprises. These same data have also been used to locate the specific enterprise markets in respect of the seasonality of arrivals and prices in the major markets, as well as the wholesale and retail prices of different commodities specific to Soroti. These unpublished data were gathered by visiting the commercial and marketing department of the Soroti District local government.

Primary data were collected through fieldwork in Soroti District with the Department of Agriculture, the Department of Commercial Services, individual farmers, farmers' associations and groups, and traders in grain. This involved the participatory generation of information through interviews and focus-group discussions. Fieldwork consisted of interviews with 50 stratified and randomly selected households, three focus-group discussions involving more than 60 farmers, and meetings and discussions with more than ten produce dealers. Semi-structured interview tools and interview guides were used to elicit experiences with climate change and food availability from the different stakeholders. In addition, five in-depth interviews with key informants from Soroti District, including the District Commercial Officer, District Agricultural Officer, three non-governmental organization (NGO) staff from World Vision and the Soroti Rural Development Agency were conducted. Qualitative and quantitative field interviews were carried out with farmers involved in the production of specific grains and other enterprises of interest.

Neoliberalism, climate change and producer price volatility in East Africa

> Neoliberalism is a theory of political economic practices that proposes that human well-being can best be advanced by liberating individual entrepreneurial freedoms and skills within an institutional framework characterized by strong private property rights, free markets, and free trade.
>
> (Harvey, 2005)

During the period from independence up until the beginning of the liberalization process in the 1980s and 1990s, agricultural markets in sub-Saharan Africa (SSA) were controlled by the state. The national public marketing board system was common throughout SSA, where the state among other things regulated pan-territorial and pan-seasonal pricing, especially for grains. Private-sector involvement was very limited at the time, but with the introduction of liberalization and structural adjustment programmes, reforms were implemented that followed three main trajectories: (i) rollback of the state; (ii) progressive disengagement from purchasing, storing and processing and (where applicable) exporting; and (iii) liberalization of private markets (Friis-Hansen, 2000).

While reducing the role of the state and disengaging from purchasing, storing and processing were widely implemented for food crops in the whole of SSA during structural adjustment in the 1990s, the pace of the establishment of private competitive markets was slow and incomplete. The expectation was that the liberalization of agricultural markets would improve price incentives for farmers, reduce government intervention and generate a sector of well-functioning competitive private traders. Although the elimination of state and parastatal intermediary organizations to some extent increased cost effectiveness, it left a number of areas uncovered because private entrepreneurs found the returns too low or the risks too high in the absence of institutional or infrastructural support (Friis-Hansen, 2000; Kherallah *et al.*, 2000).

Because neoliberalism appeared more or less simultaneously with globalization, the geographical scope of the new economic thinking reached the most remote and less developed areas of the world. This claim contradicts many popular discourses and portrayals of developing countries as deprived and standing outside the growth experienced in many other countries around the globe. In fact, many African economies are connected to the global economy through agrofood commodities and minerals (Gibbon and Ponte, 2005), being characterized by a high number of small and vulnerable economies. This vulnerability is connected to many things, such as general agricultural livelihoods and externalities that affect crops, but also a range of unstable and largely unrewarding engagements with broader market defines African agrarian livelihoods. Gibbon and Ponte (2005) state that the prices of primary exports that many African states rely on have been falling and/or become unstable because of the volatility of international market prices. Furthermore, the value of exported products is often created at higher levels in the value chain, such as manufacturing and processing, which often takes place in other countries. A relatively lower value goes to the local producers and national economies. Thus, although African countries are part of the global economy, the benefits they reap are minimal.

Jayne and Jones (1997) argue that one of the factors in support of liberalization in east and southern Africa were the operational costs of marketing boards and pan-territorial pricing. The flow of grains from areas of surplus to areas of deficit was challenged by trade restrictions that sparked further liberalization. One of the most universal food marketing reforms was price liberalization, followed by the elimination of subsidies, the lifting of geographical trade barriers and the opening up of the domestic trade to the private sector.

The private sector has now come to dominate the production of food crops. In most SSA countries price liberalization led to a drop in prices because governments had high price levels in the 1980s to ensure food sufficiency, in particular in areas difficult to reach because of poor infrastructure. The more favourable areas benefited significantly from price liberalization. The main gainers of food market liberalization are the consumers and, since prices dropped, so did profit margins. The success of private-market entry is ascribed to the low level of entry barriers, especially in primary grain marketing, whereas investment in the construction of marketing infrastructure in remote areas remains limited.

A comprehensive fieldwork study from 2009 concludes that, from a national economic point of view, the benefits of agricultural trade liberalization for the region's consumers outweigh the costs to producers (Minot *et al.*, 2009). However, market liberalization has not benefited producers located in areas where markets are poorly accessible and non-competitive. Significant trade liberalization has been implemented in Uganda since 1992, being designed, amongst other things, to reverse and even eliminate the trade deficit through increasing export earnings.

A recent global assessment (FAO, 2011) of price volatility indicates that 'variations in prices become problematic when they are large and cannot be anticipated and, as a result, create a level of uncertainty which increases risks for producers, traders, consumers and governments and may lead to sub-optimal decisions' (FAO, 2011, p.6). Climate change events such as extreme droughts are one such situation that has contributed to unprecedented price volatility. Increased volatility in crop production results in greater volatility in seasonal producer prices. A liberalized trade in grain during droughts can cause local or regional supply problems that could enhance volatility further (FAO, 2011).

Rural households have a long tradition of coping with predictable variations in weather patterns, but addressing unpredictable changes associated with extreme climate change events such as drought are beyond the capacity of rural households. The hoarding behaviour of meso-level grain traders benefits from asymmetrical trade relations with smallholder farmers and enjoys a quasi-monopoly on the grain trade. Such liberalized but still uncompetitive grain markets reduce the availability of grain during periods of increasing demand, exerting upward price pressures and limiting the ability to local storage capacity to respond to the positive demand for food.

In theory, when a commodity becomes scarce in a fully liberalized market its price rises, which induces a drop in consumption and signals more investment in the production of that commodity. The efficiency of the price system begins to break down when economic shocks give rise to price movements that are increasingly uncertain and precipitous. When shocks surpass a certain critical size or threshold and persist at those levels, traditional policy prescriptions and coping mechanisms are likely to fail (Prakash, 2011). The detrimental impact of rising volatility on these economies rests on their structural disposition: poor infrastructure, poor supply response, incomplete markets, weak capacity to import, sovereign risk, dependence on a single dominant staple and susceptibility to climatic disturbances.

Findings

The 2013 rainfall shortage in Soroti: implications for rural food security

In order to cope with various price and production risks, farmers in developing countries normally engage in informal risk management mechanisms, ranging from income diversification activities and production strategies to common risk-sharing mechanisms based on kinship and social networks. Antonaci *et al.* (2014) suggest that these traditional risk-management methods tend to fail in the presence of larger shocks like climate change affecting wider areas. The evidence suggests that, without formal risk management, less risky and less profitable farming practices are adopted, resulting in lower productivity. By extension, the interdependence also means that adaptation to the negative impacts of climate change at the farm level can only be achieved if there are supportive policies in the face of climate risks.

Findings from the field study in Soroti indicate that farmers experienced little rainfall in the second season (August to December) of 2013. Testimonies from farmers, technical staff and traders all show that the season had few, intermittent rains that could not support crop growth. 'The rains in the second season came too late too little, and drought caught the crops at their most critical stage of flowering and forming pods. It used to rain once in a while and very little rain. We could receive one heavy downpour and spend over two weeks without rain,' a farmer in Arapai sub-county told the study team. Data from the Ugandan Meteorological Authority indicate that Soroti received 1251.6 mm of rainfall in 2013 compared to 1403.9 mm in 2012, 1414.3 mm in 2011 and 1615.3 mm in 2010. The data further indicate that Soroti received relatively less rainfall in the second half (July to December) of 2013 compared to the first half of the year (January to June): 603.9 mm in the second half of the year compared to 647.6 mm in the first half. The data further show that in the second halves of the years 2014, 2012, 2011 and 2010 there was more rain compared to the second half of 2013. The rainfall for the years 2014, 2012, 2011 and 2010 was 727.1 mm, 850.5 mm, 925.2 mm and 714.5 mm, respectively, compared to 603.5 mm in 2013. Further analysis of the data indicates that the month of August only received 134.7 mm as the highest compared to the long-term average of 207 mm (UBOS, 2013) of rain in the same month.

All the farmers interviewed for this study indicated that they rely on rain to produce their food. None of the farmers reported having used any form of irrigation. According to participants in the study, the drought in the region led to significant reductions in the local availability of food, which triggered dramatic localized increases in prices. The second season of 2013 was described as an unprecedented season of drought compared to the seasons of 2012, 2011 and 2010. In the season described as having been hit by drought (July to December 2013), data sources from the meteorological authority indicate that the region received 603.9 mm of rainfall as compared to 850.5 mm in 2012, 925.2 mm in 2011 and 714.5 mm in 2010. Moreover, there was no single month in the same period where the rainfall recorded exceeded 135 mm of rainfall, as compared to 3 months that did that were recorded for 2012 and 2011. The district production department indicated that some parts of the district have experienced their lowest rainfall in the recent past. The production department further reiterated that, because of this pattern of rainfall, some parts of the district experienced near total crop failure and in others up to 50% crop failure. 'The flowers dried out and they were not able to be pollinated, and even where they were actually pollinated, the crops literally burnt and that has caused the farmers great economic loss,' an officer from the District Production Department told the study team. The loss was not limited to the crops themselves, as pastures, an important input for livestock keepers, were also affected by drought.

'The farmers usually sell crop produce to meet household needs,' an extension worker told the team, adding, 'it means farmers have limited sources of money to meet their household needs,

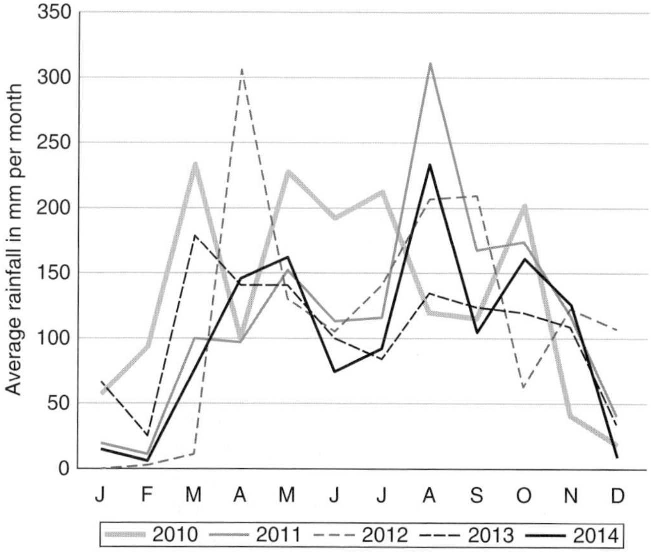

Fig. 12.1. Soroti district monthly rainfall 2010–2014. Source: Generated from Uganda Meteorological Authority data.

hence have to resort to selling livestock, which ends up depleting all their livelihood resources, hence making them more vulnerable to subsequent shocks'.

Private sector response: crop failure as a business opportunity

More than 90% of the farmers interviewed for this study indicated that either they sell their produce to traders within the first month of harvest to meet the household's immediate needs (73.2%) or because of a lack of appropriate storage facilities (64.1%), and thus farmers fear that their produce might lose value owing to deterioration. Conversely, almost every grain trader stores produce until the new stocks are due to come on to the market, which is usually every 3–4 months.

Data from the Commercial Services Department of Soroti District indicate that maize prices (wholesale and retail) following the season described by participants as an unprecedented drought period followed seasonal downward trends, despite the well below average supply of production from farmers. Maize prices during the second half of 2013 and first half of 2014 indicate that the prices of the different grains remained relatively high but stable compared to what is considered normal around the same period in other years. Prices of maize remained relatively low and stable in July and began to rise shortly in December before falling again to stabilize at an average of UGX862.5 (Ugandan Shillings) for the period between January to June, this being attributed to the flow of maize imports from the regional markets. Wholesale and retail prices for other staple grains such as beans and groundnuts began to rise as early as September 2013. The prices for beans, for example, increased atypically and did not follow the seasonal downward trend after the seasonal harvest but instead averaged UGX2400 for the next 6 months before falling to UGX1050 in June, the month of the new harvest, and then stabilizing at an average of UGX1537.5 for the rest of the year. Retail prices for groundnuts show similar trends. They started rising immediately in the drought season, reaching a maximum of UGX5250 in May before falling drastically in June to an average of UGX3400 for the rest of the year (Fig. 12.2).

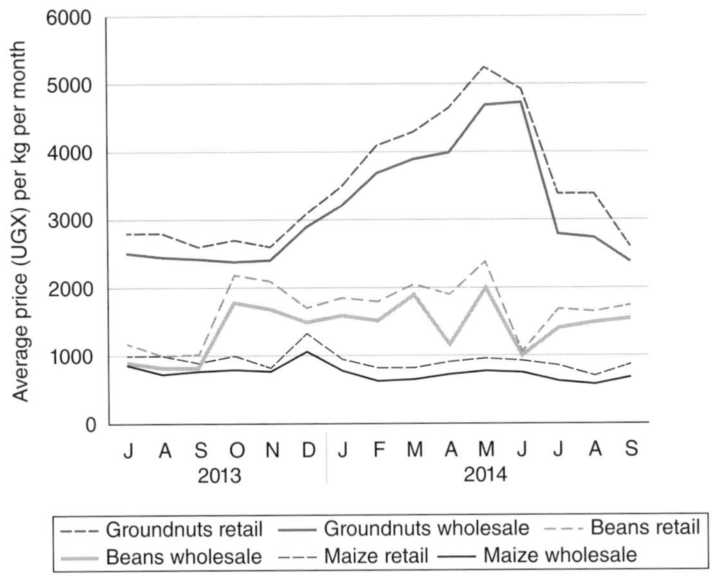

Fig. 12.2. Trends in prices of different grains over time in Soroti district before and after prolonged drought. Source: Generated from data from the Soroti District Commercial Department.

Early warning used by grain traders to refine hoarding behaviour

The study revealed that traders use early warning information and intelligence to monitor other parts of Uganda in order to determine when food surpluses will be available. They then release their stored grain when the price peaks, which is usually 2–4 weeks before grain becomes available from other parts of Uganda, which may cause the price to fall. Traders told the research team that they are always keen to establish the performance of the crops in particular seasons. 'I normally gather information on how crops are performing within this area and across the country. When the new harvests are about to arrive on the market, then I reduce the prices; but when I realize that most crops are failing due to drought, it means a very big opportunity for me to make money, and I decide to hold the produce in stock until the right time,' one trader told the study team. The study established that the right time is when the supply of produce on the market is low and demand for it is high. The study established that traders have means of gathering early warning information about rainfall shortages and the performance of different crops and behave in a way that will benefit them.

Local government perceptions of hoarding during drought

Interaction with officers from the Commercial Department of Soroti District revealed that a majority of traders stock grains for up to 4–6 months depending on availability. All farmers sampled in the household survey mentioned that they produce for subsistence and lack the capacity to construct granaries. The farmers highlighted difficulties in their ability to meet current and future food needs with their existing production capacity. Amongst the households interviewed, 83.5% of farmers said that they sell all their produce at the time of harvest, 13.4% sell 1 month after harvest and about 3% sell their produce after 2 months. The study established that farmers' capacity to store produce is hampered by difficulties in their ability to meet current and future food needs given their existing production capacity and the lack of money to construct granaries. The analysis of production capacity indicates that on average maize farmers produce 1.2 Mt/Ha compared to the national average of 1.94 Mt/Ha (Okoboi, 2010), far below the research recommendations per hectare. Groundnut production averaged 0.6 Mt/Ha and beans 1.7 Mt/Ha.

Information gathered from the district commercial office indicates that the office does not collect data on stocks held by private traders, hence most data are inferences from reported levels of production and trade. The officers viewed the practice of keeping stock for some time before prices rise as normal behaviour and consider that market forces can stabilize and manage the prices. 'We are aware of what traders are doing, but there is not much choice to do. We are operating in a liberalized economy where we cannot regulate the traders but hope the competition will lower prices as there are no restrictions in the flow of produce from one location to another. We can only advise our farmers to have food reserves.' Hoarding behaviour was not seen as abnormal by the local government marketing officer. However, the farmers who were interviewed described the traders' behaviour as unethical when the food supply in the market is low. Farmers thought that the government should formulate policies to control food prices, restrict raw food exports, expand agricultural research and promote varieties that are tolerant to drought. The farmers interviewed reiterated that the major reasons for the downward rigidity of food prices during times of food shortages is due to the imperfections in the market, the collusive behaviour of traders and asymmetrical access to information relevant to their agriculture production.

Discussion

Climate change hazards associated with extreme drought are amplified by meso-level grain traders

Uganda's government has practically eliminated price controls in the domestic market through the consistent pursuit of free-market trade and economic policies amidst concerns over climate change impacts like droughts. The absence of an autonomous and competent competition authority undermines adherence to the quest for market-determined pricing structures. Yet the strength and frequency of such extreme climate change events are likely to increase in the future (Ministry of Water and Environment, 2012). This increased risk and uncertainty is already beginning to create challenges for smallholders in the Teso sub-region who are directly impacted by extreme drought, landslides and extreme floods.

The increases in the staple grain price in the first half of 2014 caused hardships on a number of fronts. The steep rise in food prices led to economic difficulties for the poor. The price spike was triggered partly by policy failures and market overreactions. One important factor in the crisis was the unregulated hoarding behaviour of traders, which contributed to a price spike during the first 6 months of 2014. The traders kept releasing small amounts of grain on the market as they monitored new stocks, thus keeping the prices for staple grains high. This incident highlights the need to regulate traders to address the problem of price spikes, especially their effects on the livelihoods of the poor faced with climate change disasters.

Although trade liberalization policy was meant to offer products and services competitively, reliably and sustainably where there are no export restrictions on agricultural products, nor any ban on trade in food, these actions are not sufficient to avoid extreme price spikes and to ensure that smallholders can respond to emergency needs for food and seed.

Grain market prices are mainly determined by underlying conditions of supply and demand in the market (Miguel *et al.*, 2004). This implies that droughts impede production and income at the household level, which in turn affects aggregate output and welfare (Asiimwe and Mpuga, 2007). Liberalization was expected to deliver increased agricultural production and incomes to farmers and improve food security, which is not the case, however, when drought and other climate change effects start to pinch.

Adapting to climate change by regulating trade

In the pre-liberalization period, agricultural commodity prices were controlled by the state. Critics accused this of distorting the markets, leaving prices hostage to the whims of bureaucrats, bribe-taking officials and traders. While 'speculation is crucial to the proper functioning of markets, there is strong conviction that unlimited speculation is not. The central argument here is that once speculation becomes "excessive"– to the point that the marginal benefit of the liquidity that speculators provide exceeds the marginal cost of the damage that they do to the price discovery function – there is need for intervention' (Prakash, 2011). There is thus a need for a return to market management and control, ranging from the coordination of supplies – either through national or global buffer stocks, including international commodity agreements (ICAs) – to indirect market interventions via so-called 'virtual stocks' in organized exchanges. However, such proposals suffer from the fact that any policy that purports to manage the fundamentals of a commodity market cannot control the actions of the myriads of private agents that are a feature of all food markets. Moreover, speculators can normally counteract the actions of all but the best-financed interventions (Prakash, 2011).

There is an analogy here between natural disasters and food crises: clearly, any form of prevention is much costlier after a major disaster than before, but perceptions about the need for prevention are strongest only after a disaster, not before. There is thus a great need for the government to prescribe stockholding limits by issuing grain control orders depending on the basis of its assessments of market conditions. During bumper harvests, it can remove these limits. It is risky, however, to rely on the private sector to control prices in times of disaster.

As an effect of marketable liberalization, the regulatory role of governments in determining agricultural product prices has been strongly reduced. This raises the question if, when and how governments should intervene in situations of failure in the way markets operate. Market failures occur when the costs and benefits that guide individuals and the private sector differ from those that are economically optimal for society as a whole. One such situation that occurs at regular intervals is when extreme droughts and market failures result in a deficit of food grains and artificially inflated prices in vulnerable areas.

> Many of the key barriers to trade in food staples relate to regulatory and competition issues at elements along the value chain. As tariffs have come down, so a tangled web of rules, fees, and expensive services is strangling Africa's regional trade in food. In some cases, the policies restricting trade are deliberately protectionist; but in other instances, the lack of regulations limits the development of regional markets, such as in the case of fertilizers. In many cases the issue is to define appropriate regulations and create better institutions to implement them.
> (World Bank, 2012, p.18)

Conclusion

Because of the asymmetric power relations between many small-scale autonomous and unorganized farmers and a few large traders, liberalization has led to a de facto oligopoly where the traders at farm-gate level are not effectively competing but rather taking full advantage of the vulnerability facing farmers by aggressively early buying the pre-harvest. As the farmers are highly vulnerable and desperate to sell, the traders can buy at low prices. This is also a reflection of market inefficiencies and failures, as well as the weakness of the value chain at the farm gate, as transaction costs limit the number of traders in a given locality.

The effects of hoarding are enhanced by exogenous shocks and the risks imposed by climate change, namely: (i) low prices during pre-harvest buying; and (ii) high prices when farmers need food 3–4 months after the harvest and none is available on the market.

The study has demonstrated that trade liberalization policy fails to deal with climate change hazards. Trade liberalization policy focuses mainly on market distortions made by governments but ignores those that are evident in the private sector. Climate change exacerbates the problem of the already existing low agricultural output and in the worst-case scenarios there is total crop failure resulting in food insecurity. When there is crop failure, farmers resort to the market for both their food and the seed they need for the next season. When disaster strikes and there are looming price hikes, there is a need for mechanisms to be in place that enhance long-term and efficient staple grain supplies, ensure that appropriate levels are maintained in the market and promote investment in climate-smart agriculture in order to make smallholders more resilient.

References

Antonaci, L., Demeke, M. and Vezzani, A. (2014) The challenges of managing agricultural price and production risks in sub-Saharan Africa. *ESA Working Paper No. 14-09*. FAO, Rome.

Asiimwe, J.B. and Mpuga, P. (2007) Implications of rainfall shocks for household income and consumption in Uganda. RP 168, African Economic Research Consortium, Nairobi, Kenya.

FAO (2011) *The State of Food Insecurity in the World; How does International Price Volatility Affect Domestic Economies and Food Security?* FAO, Rome.

Friis-Hansen, E. (ed.) (2000) *Agricultural Policy in Africa After Adjustment*. CDR Policy Paper. Centre for Development Research, Copenhagen.

Gibbon, P. and Ponte, S. (2005) *Trading Down: Africa, Value Chains, and the Global Economy*. Temple University Press, Philadelphia, PA.

Harvey, D. (2005) *A Brief History of Neoliberalism*. Oxford University Press, Oxford, UK.

Jayne, T.S. and Jones, S. (1997) Food marketing and pricing policy in Eastern and Southern Africa: a survey. *World Development* 25, 1505–1527.

Kherallah, M., Delgado, C.L., Minot, N. and Johnson, M. (2000) *The Road Half Traveled: Agricultural Market Reform in Sub-Saharan Africa*. IFPRI, Washington DC.

Miguel, E., Satyanath, S. and Sergenti, E. (2004) Economic shocks and civil conflict: an instrumental variables approach. *Journal of Political Economy* 112, 725–753.

Ministry of Tourism, Trade and Industry (2007) *National Trade Policy*. Government of Uganda, Kampala, Uganda.
Ministry of Water and Environment (2012) *Uganda National Climate Change Draft Costed Implementation Strategy*. Ministry of Water and Environment, Government of Uganda, Kampala, Uganda.
Minot, N., Chemingui, M.A., Thomas, M., Dewina, R. and Orden, D. (2009) *Trade Liberalization and Poverty in the Middle East and North Africa*. International Food Policy Research Institute (IFPRI), Washington, DC.
Okoboi, G. (2010) *Improved Inputs Use and Productivity in Uganda's Maize Sub-sector*. Research series no. 69. Economic Policy Research Centre, Kampala, Uganda.
Prakash, A. (ed.) (2011) *Safeguarding Food Security in Volatile Global Markets*. FAO, Rome, Italy.
UBOS (2013) *Statistical Abstract*. Uganda Bureau of Statistics, Kampala, Uganda.
USAID (2013) Uganda Climate Change Vulnerability Assessment Report. Available at https://www.climatelinks.org/resources/uganda-climate-change-vulnerability-assessment-report (accessed 9 June 2017).
World Bank (2006) *Uganda Diagnostic Trade Integration Study*. World Bank, Washington, DC.
World Bank (2012) *Africa Can Help Feed Africa. Removing Barriers to Regional Trade in Food Staples*. Poverty Reduction and Economic Management, Africa Region, Washington, DC.

13 Conclusion: Towards Subnational Climate Change Action Amidst Ambiguity and Contestation Over Power and Resources

Esbern Friis-Hansen

Governance of Climate Change Action

Moving from the COP 21 agreement and National Climate Change Policies to action at the subnational and community levels involves more than just capacity building and finance. Although these are important issues, they cannot be addressed or understood outside the context of governance. Ambiguity over how best to adapt to climate change hazards is common among all types of actors at all levels of government. Adapting to climate change hazards in rural areas is complex because there are no easy adaptive responses to increased variability in weather patterns and the occurrence of extreme events such as devastating floods and droughts. Individual attempts to cope with the situation are inadequate, and effective climate change adaptation calls for collective action.

As a consequence of the centralized governance structure of climate change policy as well as a delay in access to international finance due to the prolonged COP negotiations, international and national support for climate change adaptation in rural areas has until now been limited compared with that in urban centres. The need of the rural population in Africa to create an environment that enables them to adapt to climate change has largely been ignored. Meanwhile international and African government actors at all levels use the ambiguity associated with climate change as a new political arena for contesting existing power relationships.

Climate change action by both national and subnational levels of governance is used as a means of gaining control over access to funds and decision-making power. Through the ministries responsible for climate change, the national level of government has dominated the policy formulation process in conjunction with international development partners. Chapter 2 show that, while there were considerable participatory consultations during the formulation of National Climate Change Policies (NCCPs) in Uganda, the governance process for implementing climate change action remained centralized. An analysis of the 11 available NCCPs for Africa indicates that, while these policies pay lip-service to existing decentralization policies and express the intention of working through them, they implement a de-facto de-concentration approach to climate change action that is non-compliant with decentralization legislation.

The implications of the centralization of decision making and control over funding by national climate change policies are many. Most importantly, centralization creates a disconnect between where decisions are made and those layers of government where climate change action happens. Subsidiarity principles of the governance of climate change adaptation are thus ignored, and local governments are not encouraged to create a political space for the participatory planning processes that are needed to mobilize local citizens. Climate change adaptation is highly discretionary and to be effective it requires people's direct involvement in decision making and control of resources; however, centralized control of funding and decision making for climate change adaptation disempowers local people.

The disconnect between central-level policies and local needs means that access to financial resources is skewed and local governments are unable to direct the technical inputs of local-sector agencies according to locally developed adaptation needs and plans. Frustration with this is growing among technical staff at both central and local government levels. As Chapter 5 argued, this may potentially constitute a common future meeting ground between local governments and de-concentrated technical staff (in the case of Zambia) and with technical staff under devolution, but reduced to implementing agents by centrally managed projects (in the case of Uganda).

Climate Change Action as an Arena for Political Contestation

The centralization of fiscal resources and political power also changes the roles of and relationships between central ministries and local government. Rather than being political and administrative partners, local governments are assigned a role of implementing agents by central government and international development agencies. However, rather than surrender their authority to central government, local governments work actively to consolidate and extend the possibilities available to them. With examples from climate change and other adaptation-related areas such as agriculture and water, Chapter 4 showed how two district councils in Zambia manoeuvre pragmatically between associating themselves with the state and distancing themselves from it and claiming autonomy. Using three different strategies, Zambian district councillors employ climate change as a platform to challenge the authority, efficiency and control of resources of the central state. Firstly, by openly criticizing state agencies for failing to address the impacts of climate change adequately, district councillors are making it difficult for the central state to disregard councils and arguing for the existence of alternative alliances beyond the central state. Secondly, councils in the two districts employ climate change in discourses of devolution. During engagements with central government and development partner agencies, local government employees and politicians argue that, based on principles of subsidiarity, they are the best-placed organization to address climate change. Thirdly, climate change is used by local government politicians as a platform for challenging the central state's control over land and other natural resources. Although district councils in Zambia disapprove of their limited mandate and inadequate financing for climate change adaptation, they engage actively in adaptation efforts whenever opportunities allow, such as harnessing the resources and know-how of non-governmental organizations (NGOs). District councils actively seek to enforce a requirement that NGOs consult and plan all activities in close collaboration with themselves, rather than implementing their projects independently.

Social processes and struggles over authority also take place as part of the implementation of climate change strategies at the subnational level. Chapter 3 examined local struggles within local government departments in Tanzania in the area of water resources. Technical staff in the Rufiji Basin Water Office and the district departments of Planning, Community Development and Land, and Forests and Natural Resources are aware of recent climate policies such as the National Climate Change Strategy and the REDD strategy and are actively considering their mandate in the area of adaptation implementation in terms of regulating the utilization of water resources. Meanwhile the Department of Agriculture did not refer to climate change policy during interviews but instead was promoting the further expansion of irrigation for crop production. Chapter 3 showed how climate change adaptation planning and implementation by individual district departments and the Rufiji Basin Water Office have different degrees of upward and downward accountability and different amounts of leeway depending on their national-level financial, political and institutional backing, and therefore different allegiances and feelings of responsibility.

This has major implications for adaptation planning and implementation on the ground. Most district revenue comes from centrally transferred sectoral block grants, from Local Capital Development Grants and to a lesser extent from direct donor support. As a consequence of the centralized governance of climate change, the district has no official budget lines for climate-related spending and only limited budgetary discretion within the agricultural domain. The agricultural department is strongly influenced by central government guidelines to prioritize activities aimed at poverty reduction, socio-economic development and food security through agricultural production-led growth and less concerned with climate change adaptation. District departments for Planning, Community Development and Natural Resources are constantly looking for opportunities to break down the hegemony of agriculture. With only a fraction of the agricultural resources at their disposal, these departments have a track record in making the most of limited resources.

When Local Government is Disempowered, the Institutional Sustainability of Support for Climate Change Adaptation Suffers

Chapter 5 analysed the Territorial Approach to Climate Change (TACC) in Uganda as an example of a project approach to supporting rural climate change adaptation that is widely used in different variations by central government and international donor agencies. The aim of the analysis was not to evaluate TACC but to seek to understand it as an approach to governance and to explore its implications for how internationally financed support engages with subnational government. A key lesson learned from the TACC study is that implementing climate change adaptation support through project structures at the subnational level that are parallel to those of district local government disempowers locally elected political leaders in the district council. In spite of a considerable level of popular consultation and leaders being invited to take part in the steering committee that prioritizes support for local NGOs and community-based organizations (CBOs), such instrumental participation remains a superficial varnish that does not hide the lack of real devolution of control over financial resources and decision making.

An associated conclusion that echoes other parts of the book is that local politics matters. When district politicians are disempowered from direct control over the budget and only play a consultative role in priority setting, they do not have a sense of ownership of the project, however technically sound it may be. Without the support of local political leaders, the project will not be integrated into the district development plan, and project activities will discontinue when finance comes to an end. Effective climate change adaptation requires an ongoing localized inclusive process involving all stakeholders and supported by legitimate political leaders. Although the project approach and use of international and national consultants ensure effective implementation enabling the project to honour the outcomes indicated in its log frame, it fails to sustain its activities after project closure.

One should not underestimate the effect on local government technical staff of the extremely skewed use of fiscal resources associated with implementing climate change support using structures that are parallel to local government administration. The enormous differences in access to resources for climate change action between the project staff and local government technical staff not only reduced the role played by local government technical staff in formulating the Integrated Territorial Climate Plan from the status of contributing partners to passive agents, it also meant that very little capacity development among local government staff took place during TACC implementation.

Finally, the TACC study indicates that, when interpreting the aspirations of rural people affected by climate change hazards is undertaken by external consultants and project management, rather than through a localized inclusive political process, the recommendations will be biased towards technical and infrastructural solutions. Aside from their high cost, a fundamental problem associated with such technical solutions is they are not based on a broadly accepted political compromise and are thus likely to be challenged during implementation.

NGO Facilitation of Interaction Between Local Government and Communities

Because climate change affects each community differently, and needs and vulnerabilities vary greatly even within very small geographical areas, acknowledging communities' claims to participation becomes vital in developing effective adaptation plans. Individual communities are best at identifying their needs and have different adaptive capacities, meaning that, conversely, applying best-practice solutions to several communities can result in lower results with the same or perhaps even higher costs. Chapter 6 showed how CARE, an international NGO, has facilitated a process of reshaping as well as creating a political space between local governance and local communities by inviting both actors into the process. The study shows that NGOs can play a significant role in reconfiguring the political space between local government and communities and in creating new spaces by unfolding participatory approaches integrated in and received by local communities.

Successful local governance of climate change adaptation depends on the existence of sustainable multi-level and cross-scale governance structures. By focusing on extending invitations to participate, while at the same time strengthening the community's ability to claim a political space in its own right, the NGO facilitates engagement by both local communities and local government in the process of strengthening participation, thus making climate change adaptation a collective process. This approach can be understood as involving pragmatic attempts to make participation more transformative and empowering by shifting the focus beyond the individual and local levels to involve multi-scale strategies that encompass the institutional and structural aspects of governance. An emphasis on cross-scale governance with a special focus on the local scales is shown to be influential in improving adaptive climate change governance if scale governance is effectively integrated with multi-level governance.

Although NGO facilitation of invited political spaces may allow local communities a voice in policy-making processes and possibly influence the agenda and rules of engagement, it does not guarantee that the process will foster a longer lasting critical consciousness and inclusive decision making, or whether it remains a methodology for improving project performance. The study emphasized the need to pay attention to the fact that such participatory development processes are created by formal agencies of development flowing from us to them, and that success depends upon the knowledge, institutions and best practices of professionals.

Overall, however, opening downward accountable political spaces for dialogue strengthens the overall decentralization process because the approach forges greatly needed downward accountability links between local communities and local government. Both communities and local governance highlight the lack of government funding as the main problem for further development. It has, however, paved the way for a more effective pathway for future flow of international funding for climate change adaptation.

Climate Change on the Political Agenda of Local Government: a Lose–Lose Issue?

Chapter 9 was concerned with how adaptation to climate change is articulated politically at the subnational level, especially among the very many rural local governments that do not yet receive any financial support from central government for climate change action. The study revealed that local politicians perceive this as a 'lose–lose issue'. First, the absence of a budget line for climate change adaptation has contributed to many local government politicians being inactive, the general trend being that, without a budget line, climate change adaptation is not on the district council's political agenda: 'No funding – no activities', as one politician put it. Secondly, the challenge of climate change is perceived by local politicians as complex and challenging to solve because it involves adaptive natural resource management solutions that may be beneficial and lower risk for many people, but is likely to aggravate others (many of whom are among the local elite) as their unsustainable economic activities may be regulated. 'There are no votes in climate change adaptation', as another politician noted.

However, by studying an atypical district in which strong climate action had been taken, the study showed that it is possible to put climate change on the political agenda of local government even without any additional budget. In the case of Amuria District in Uganda, one charismatic opposition politician won the popular vote during the 2006 elections to the District Council, using the widespread concerns over climate change among the electorate as part of his political platform. After his election victory, the new district council chairman insisted on acting on environmental issues, including climate change, in spite of the absence of a budget, by ordering the environmental officer to develop a participatory plan of action. The formulation and enactment of an environmental ordinance and the formulation and implementation of Wetlands Management Plans (WMPs) proved to be a viable policy instrument for local government. It created what can be understood as an enabling environment for climate change adaptation against extreme climate change events in the form of floods.

As no additional budget was available, all stakeholders in the formulation of the WMP had to ensure that the plans were legitimate in the eyes of the majority of the rural population whom the district relied upon for its enforcement. The political champion provided the necessary political clout to the process, making it possible to disempower the local elites involved with rice cultivation and to make plans to protect further encroachment in the wetlands and ensure sustainable plan use practices. The participatory planning process can be seen as giving a voice to the new actors in wetlands management, that is, giving greater space to women and poor farmers, who, under the old regime, were threatened by commercial rice production. However, the study also shows that enhanced invited political space and downwardly accountable institutions alone cannot empower people. The absence of financial resources and the low level of political interest and support from local government after the district chairman was replaced at the 2011 election has reduced the effectiveness of the new institutions for enforcing the restrictions on encroaching of wetlands.

The Political Ecology of Climate Change: Processes of Institutional Bricolage Where Government is Weak

Section 3 focused on political ecology and climate change adaptation at the subnational level in the context of a rural population affected by climate change hazards, but where at the same time state institutions are weak at the subdistrict level in terms of their lack of both human and fiscal resources. Governance of climate change action at this level is constrained by inadequate human resources and national budgetary support under the decentralized local governance framework. Chapters 8, 9 and 10 showed how the weak presence of the state encourages or forces front-line bureaucrats to accept a collaborative rather than a leading role, as other forces claim political space. Local groups organized around an enterprise exercise power in a variety of arenas and at multiple scales, thereby influencing the implementation of regulations for the protection of natural resources.

The governance of natural resources in a changing climate is embedded in political, economic and social processes at the district and sub-county levels. The changing political processes, social issues, economic issues and how they interact influence stakeholders' engagement, flexibility in implementing natural resource regulations and resource access and use. The three chapters examined different aspects of environmental change in Awoja, showing how they are embedded in the struggle of community members to adapt to the effects and economic opportunities presented by a changing climate. This creates a dilemma for governance systems in choosing between promoting development and community economic survival versus adapting to extreme floods through conservation of the wetlands. Different agencies are working at different scales and with vastly different concentrations of power, from the level of household livelihoods through to the decisions of bureaucrats under the decentralization framework.

Chapter 9 examined how local government's ability to govern is not only limited by weak manpower and fiscal resources, but also constrained by the multiple accountability of its front-line staff that needs to be accountable both upwards to central government and downwards to the community, as well as sideways to their relatives and kin. Stakeholder participation in local natural resource management committees has created a space for innovations and the renegotiation of rules and regulations for the governance of natural resources under extreme climatic conditions. NGOs and CBOs have initiated many innovative measures for the management of the watershed. The legitimacy of NGOs and CBOs as advocates of sustainable resources has increased at the expense of local government regulatory authorities.

Politics of Climate Change: Commercial Enterprise Development and Elite Capture of Local Government Environmental Institutions

The commoditization of rice production in the Awoja watershed illustrates how protecting the environment and adaptation to climate-induced floods are in competition with commercial and political interests, resulting in elite capture, local government inaction and compromises in governing the watershed. Chapter 10 examined how commercial rice production in the Gweri wetlands is influencing institutional arrangements designed to govern the wetlands and leading to conflicts of interpretation over land-use management regulations. The expansion of commercial rice production is being driven by its potential to generate higher levels of income for communities than any alternative enterprises. Rice was originally promoted by a JICA-sponsored government programme and was intended for uplands as an attempt to move people way from the swamps. It did exactly the opposite, however, because farmers preferred to cultivate rice in the swamps rather than the infertile and dry uplands, and the so-called upland rice is now growing on the fringes of the swamp, causing siltation of the wetlands. Rice production is not profitable on a small scale; instead it is dominated by tractor farmers, who are in part sponsored by urban-based businessmen linked to the value chain.

The expansion of commercial rice cultivation into the swamps is creating land-use conflicts with cattle herders, who for generations relied on the wetlands as communal grazing grounds where cattle roamed without restriction. Local-level government bureaucrats are divided between enforcing environmental policies and local bylaws that seek to protect both the wetlands themselves and the commercial interests of the local elite involved in commercial rice cultivation in them. The study uncovered poor implementation of existing natural resource management policies and regulations owing to elite capture of local government institutions because Disaster Risk Reduction committee members were directly involved in cultivating rice in the wetlands and were unwilling to restrict their lucrative enterprise.

Emerging Themes for Climate Change Governance: Mobility and Migration

The fourth section explores two new and current emerging themes for climate change governance: mobility and trade. Both are important phenomena that interact with climate change, though both have been empirically under-researched in Africa.

Chapter 11 examined the changing mobility patterns of Karamajong pastoralists and their interaction with the Teso agriculturalists in Uganda under the influence of climate-induced droughts and floods. The chapter reviewed the literature and shows how international understandings of the relationship between mobility and climate change have changed over time from understanding climate change as a direct cause of migration to a general consensus today that mobility and migration have multiple causes, including climate change. In recent years, the debate has been distorted by politically motivated arguments over 'environmental refugees' that elide two separate phenomena, namely migrants and refugees. The NGO proponents who argue for the existence of climate refugees (and the need for the international community to increase funding to address climate change) revert to a simplistic understanding that the environment is the direct cause of migration. This underestimates the importance of historical migration patterns in high-risk environments, as well as the causes and processes of adaptation, and is largely at odds with empirical case studies showing the multi-causal character of migration patterns. The notion of 'environmental refugees' also wrongly portrays climate refugees as passive victims rather than active citizens.

The study showed that mobility and migration associated with climate change is largely local or regional and seldom involves cross-border migration. The study also concluded that ethnicity has a strong influence on the level of conflict associated with mobility. Karamojong who move internally from dry areas to the green belts within the Karamoja region on the border with the Teso region are not seen as a challenge because the major resource, namely land, is communally owned by the Karamajong tribe. Migrants are looked on as brothers and sisters, and as long as they comply with the community's norms, they are easily integrated. However, when Karamojong migrants settle across the border within the Teso region, multiple conflicts between the indigenous Teso people and the migrants arise immediately owing to the long history of cattle raids and armed attacks.

The surprising finding from the study is not that the settlement of climate migrants within another tribe causes conflicts over access to and use of natural resources, but how these conflicts are solved. Informal political and social dialogue has helped the two communities forge new relationships that contribute to the long-term adaptation to droughts, which have been the main cause of conflict in this cross-border area. Changes in the form of institutional bricolage emerge from new socio-economic and socio-cultural forms of interaction between the two ethnic communities, with implications for how resources are shared between the immigrants and the local population.

Emerging Themes for Climate Change Governance: Regulating Agricultural Trade

Regional food trade is rarely discussed in connection with climate change. Chapter 12 thus broke new ground by examining how climate change hazards associated with extreme drought are amplified by meso-level grain traders in Uganda. The study showed that, because of the asymmetric power relations between many small-scale autonomous and unorganized farmers and a few large traders, liberalization has led to a de facto oligopoly. During severe drought, traders at the farm-gate level are effectively not competing but rather taking full advantage of the vulnerability facing farmers by aggressively buying early pre-harvest. Because the farmers are highly vulnerable and desperate to sell during a drought, the traders can buy at low prices.

The study demonstrated the failure of trade liberalization policy to deal with climate change hazards. Trade liberalization policy focuses mainly on market distortions made by governments but it ignores those that are evident in the private sector. Climate change exacerbates the problem of the already existing low agricultural output. When disaster strikes and there are looming price hikes, there is a need for mechanisms to be in place to enhance long-term and efficient staple grain supplies, ensure that appropriate levels are maintained in the market, and promote investment in climate-smart agriculture in order to make smallholders more resilient.

As an effect of marketable liberalization, the regulatory role of governments in determining prices for agricultural products has been strongly reduced. This raises the questions of whether, when and how governments should intervene in situations of failure in the way markets operate. Market failures occur when the costs and benefits that guide individuals and the private sector differ from those that are economically optimal for society as a whole. One such situation that occurs at regular intervals is when extreme droughts and market failures result in a deficit of food grains and artificially inflated prices in vulnerable areas.

Towards a Model of Engaging Subnational Government in Creating an Enabling Environment for Rural Climate Change Adaptation in Africa

A first step towards creating an enabling environment for climate change adaptation might be to examine how subnational processes of governance and decentralization influence climate action in the given setting and to determine what needs to be done on this basis. Chapter 2 discussed how central governments often do not follow through with plans for devolution, and in some countries there is even a tendency towards the recentralization of service provision. Chapters 2, 3 and 4 discussed how national governments and their sector agencies seek to monopolize climate action, rather than relinquish control to local governments and other local actors, despite a general realization of the need to engage with the subnational government level and of the importance of decentralization policies being in place.

Climate agreements and financing mechanisms such as the Green Climate Fund are currently not engaging with subnational levels of government in a constructive manner, as they are embracing centralized projects as a mode of implementation. They might, however, play a role in this respect by incorporating measures that enhance the downward accountability of national governments and support platforms for local governments, civil society and other actors to counter autocratic adaptation (Funder, 2016).

In 2017 the World Bank, which has invested greatly in the decentralization reforms, will initiate what is likely to be one of the most serious engagements by international development agencies with the subnational government level in the form of a countrywide climate change adaptation programme in Kenya that will promote institutional arrangements to allow subdistrict level institutions to control part of the budget and to hold local government service providers downwardly accountable. This World Bank-funded loan programme combines the Community Based Development approach with the devolution of climate change funding to the subdistrict level (World Bank, 2016).

One of the arguments for continuing the project approach to supporting rural climate change action is a concern for whether there is adequate human capacity at the local government level to manage large investments. However, while there is little doubt that there is a need to strengthen the human capacity to address climate change at the local government level, this book argues that the key to providing capacity support is a matter of governance. That is, local governments should be in a position to direct the technical inputs of local-sector agencies according to locally developed adaptation needs and plans.

As discussed in the case of Zambia, local government is continually contesting the current hegemony of central government over climate change action. Frustrations over the disconnect between central-level policies and local needs is potentially a common future meeting ground between local governments and deconcentrated technical staff. Cost-effective climate change adaptation requires the devolution of political power to local government, as well as institutional mechanisms for formula-based transfers of fiscal resources as conditional grants (Friis-Hansen, 2015).

The Zambian case study in Chapter 4 also shows that climate change adaptation can provide a platform for local governments to assert their roles and claims vis-à-vis central governments more broadly. This includes forming new alliances between local governments and civil-society organizations, as well as staking claims for fiscal devolution and locally anchored natural resource governance. Chapter 6 provided an example of how such an alliance can be actively pursued from the side of an NGO. A key lesson from the Ghana study is that the structure of both the financing and governance of climate change adaptation needs to be revisited to determine how climate change governance and finance can be better integrated into the existing decentralization reform process.

References

Funder, M. 2016. Policy-makers and climate financing in African LDCs. The case of Zambia. In Adam Moe Fejerskov, Mikkel Funder, Lars Engberg-Pedersen, Yang Jiang, Helle Munk Ravnborg & Neil Anthony Webster 2016. Financing the Sustainable Development Goals. The changing actors, interests and politics in development financing. *DIIS Report*, 2016:01

Friis-Hansen, E., Bashaasha, B., Aben, C., Okiror, J.J. and Suubi, G. 2015. Local government engagement with climate change adaptation in Uganda. *DIIS Report*, 2015:19.

World Bank 2017. Kenya Climate-Smart Agriculture Project. Appraisal. http://documents.worldbank.org/curated/en/440241486868444705/pdf/Kenya-PAD-01182017.pdf

Index

Note: bold page numbers indicate tables and figures.